OXFORD MATHEMATICAL MONOGRAPHS

Series Editors
E. M. FRIEDLANDER   I. G. MACDONALD
L. NIRENBERG   R. PENROSE   J. T. STUART

OXFORD MATHEMATICAL MONOGRAPHS

A. Belleni-Morante: *Applied semigroups and evolution equations*
I.G. Macdonald: *Symmetric functions and Hall polynomials*
J.W.P. Hirschfeld: *Projective geometries over finite fields*
N. Woodhouse: *Geometric quantization*
A.M. Arthurs: *Complementary variational principles* Second edition
P.L. Bhatnagar: *Nonlinear waves in one-dimensional dispersive systems*
N. Aronszajn, T.M. Creese, and L.J. Lipkin: *Polyharmonic functions*
J.A. Goldstein: *Semigroups of linear operators*
M. Rosenblum and J. Rovnyak: *Hardy classes and operator theory*
J.W.P. Hirschfeld: *Finite projective spaces of three dimensions*
K. Iwasawa: *Local class field theory*
A. Pressley and G. Segal: *Loop groups*
J.C. Lennox and S.E. Stonehewer: *Subnormal subgroups of groups*
D.E. Edmunds and W.D. Evans: *Spectral theory and differential operators*
Wang Jianhua: *The theory of games*
S. Omatu and J.H. Seinfeld: *Distributed parameter systems: theory and applications*
D. Holt and W. Plesken: *Perfect groups*
J. Hilgert, K.H. Hofmann, and J.D. Lawson: *Lie groups, convex cones, and semigroups*
S. Dineen: *The Schwarz lemma*
B. Dwork: *Generalized hypergeometric functions*
R.J. Baston and M.G. Eastwood: *The Penrose transform: its interaction with representation theory*
S.K. Donaldson and P.B. Kronheimer: *The geometry of four-manifolds*
T. Petrie and J. Randall: *Connections, definite forms, and four-manifolds*
R. Henstock: *The general theory of integration*
D.W. Robinson: *Elliptic operators and Lie groups*
A.G. Werschulz: *The computational complexity of differential and integral equations*
J.B. Griffiths: *Colliding plane waves in general relativity*
P.N. Hoffman and J.F. Humphreys: *Projective representations of the symmetric groups*
I. Györi and G. Ladas: *The oscillation theory of delay differential equations*
J. Heinonen, T. Kilpelainen, and O. Martio: *Non-linear potential theory*
B. Amberg, S. Franciosi, and F. de Giovanni: *Products of groups*
M.E. Gurtin: *Thermomechanics of evolving phase boundaries in the plane*
I. Ionescu and M. Sofonea: *Functional and numerical methods in viscoplasticity*
U. Crenander: *General pattern theory*

Thermomechanics of Evolving Phase Boundaries in the Plane

MORTON E. GURTIN

Department of Mathematics
Carnegie Mellon University
Pittsburgh, PA, USA

CLARENDON PRESS · OXFORD
1993

Oxford University Press, Walton Street, Oxford OX2 6DP
Oxford   New York   Toronto
Delhi   Bombay   Calcutta   Madras   Karachi
Kuala Lumpur   Singapore   Hong Kong   Tokyo
Nairobi   Dar es Salaam   Cape Town
Melbourne   Auckland   Madrid
and associated companies in
Berlin   Ibadan

Oxford is a trade mark of Oxford University Press

Published in the United States
by Oxford University Press Inc., New York

© M. E. Gurtin, 1993

All rights reserved. No part of this publication may be
reproduced, stored in a retrieval system, or transmitted, in any
form or by any means, without the prior permission in writing of Oxford
University Press. Within the UK, exceptions are allowed in respect of any
fair dealing for the purpose of research or private study, or criticism or
review, as permitted under the Copyright, Designs and Patents Act, 1988, or
in the case of reprographic reproduction in accordance with the terms of
licences issued by the Copyright Licensing Agency. Enquiries concerning
reproduction outside those terms and in other countries should be sent to
the Rights Department, Oxford University Press, at the address above.

A catalogue record for this book is available from the British Library

Library of Congress Cataloging in Publication Data
Gurtin, Morton E.
Thermomechanics of evolving phase boundaries in the plane / Morton E. Gurtin
(Oxford mathematical monographs)
Includes bibliographical references (p.   ) and index.
1. Thermodynamics.   2. Phase transitions.   3. Crystal growth.
I. Title.   II. Series.
QC311.G985   1993   536'.7—dc20   92-38016
ISBN 0 19 853694 1

Typeset by Integral Typesetting, Great Yarmouth, Norfolk
Printed in Great Britain by
Biddles Ltd, Guildford & King's Lynn

For
Peg

PREFACE

This book is based on a course given at Carnegie Mellon and on a lecture series given at the Summer School on Mathematical Physics in Ravello, Italy; the Istituto Mauro Picone in Rome; and the Institute for Mathematics and its Applications, University of Minnesota.

The book is not meant to be comprehensive; it presents topics that have interested me over the past few years, and it represents a point of view different from that prevalent in the physical literature. Being an engineer by training, but a mathematician by choice, I have tried to make the book comprehensible to both mathematicians and physical scientists. Throughout I emphasize issues that are foundational in nature, as I am more interested in the interplay between mathematics and physics than in the solution of specific problems. I present what I hope are rational derivations of those free-boundary problems that form the basis of the subject. These problems should be of great interest to analysts. Unfortunately, technical mathematical issues such as existence, uniqueness, and regularity of solutions are discussed only superficially; these are better left to mathematicians more competent in this area than I.

Throughout the development of this material I have profited greatly from discussions with Sigurd Angenent, Paolo Podio-Guidugli, and David Kinderlehrer. I also acknowledge valuable discussions with Perry Leo, Jose Matias, William Mullins, Robert Sekerka, Mete Soner, Allan Struthers, Peter Voorhees, and William Williams.

My research as presented here was supported by the Army Research Office and the National Science Foundation; this support is gratefully acknowledged.

Pittsburgh   M.E.G.
September 1992

CONTENTS

0 Introduction	1
I Kinematics	**5**
1 Curves	6
1.1 Preliminary definitions	6
1.2 Convex curves	7
1.3 Integrals	10
1.4 Piecewise-smooth curves	10
1.5 Infinitesimally wrinkled curves	12
2 Evolving curves	14
2.1 Definitions	14
2.2 Transport identities	15
2.3 Integral identities	17
2.4 Steadily evolving interfaces	19
2.5 Piecewise-smooth evolving curves	20
2.6 Variational lemmas	24
3 Phase regions, control volumes, and inflows	26
3.1 Phase regions and control volumes	26
3.2 Inflows, the pillbox lemma, and infinitesimally thin evolving control volumes	26
II Mechanical theory of interfacial evolution	**31**
4 Balance of forces; power	32
4.1 Balance of forces	32
4.2 The power identity	33
5 Energetics and the dissipation inequality	35
6 Constitutive theory	38
6.1 Constitutive equations and the compatibility theorem	38
6.2 Balance of capillary forces revisited; corners	40
7 Digression: statical theory of interfacial stability; convexity, the Frank diagram, and corners; Wulff regions	41
7.1 Preliminaries; polar diagrams	41
7.2 Convexity; the extended and convexified energies, and the Frank diagram	42
7.3 Stability	47
7.4 Instability of the total energy	51
7.5 Equilibria of the total energy; Wulff regions	55
7.6 Wulff's theorem	57

CONTENTS

8 Evolution equations for the interface; basic assumptions ... 59
   8.1 Isotropic interface ... 59
   8.2 Anisotropic interface ... 60
      8.2.1 Basic equations ... 60
      8.2.2 Equations when the interface is the graph of a function ... 61
      8.2.3 Equations when the interface is a level set ... 62
   8.3 Plan of the next few chapters ... 63

9 Stationary interfaces and steadily evolving interfaces ... 64
   9.1 Stationary interfaces ... 64
   9.2 Steadily evolving facets ... 65
   9.3 Steadily evolving interfaces that are not flat ... 65

10 Global behaviour for an interface with stable energy ... 69
   10.1 Existence of evolving interfaces from a prescribed initial curve ... 69
   10.2 Growth and decay of the interface ... 70
   10.3 Evolution of curvature; fingers ... 73

11 Unstable interfacial energies and interfaces with corners ... 78
   11.1 Admissibility; corner conditions ... 78
   11.2 The initial-value problem ... 80
   11.3 Facets and wrinklings that connect evolving curves ... 81
   11.4 Equations near a corner when the curve is a graph ... 84
   11.5 Interfaces with arbitrary angle-set; infinitesimal wrinklings ... 85
   11.6 Stationary interfaces and steadily evolving interfaces with corners ... 90

12 Non-smooth interfacial energies; crystalline energies ... 91
   12.1 Crystalline energies ... 91
   12.2 The Wulff region ... 92
   12.3 The capillary force at preferred orientations ... 93
   12.4 Corners between preferred facets ... 94
   12.5 Crystalline motions ... 96
   12.6 Interfaces of arbitrary orientation, infinitesimal wrinklings, and generalized motions ... 100
   12.7 Evolution of a rectangular crystal ... 103

13 Regularized theory for smooth unstable energies; dependence of interfacial energy on curvature ... 105
   13.1 Balance of forces and moments; power ... 105
   13.2 Energetics and the dissipation inequality ... 106
   13.3 Constitutive equations ... 107
   13.4 Evolution equations for the interface ... 109
   13.5 Linearized equations; spinodal decomposition on the interface ... 110

III Thermodynamical theory of interfacial evolution in the presence of bulk heat conduction ... 111

14 Review of single-phase thermodynamics ... 112
   14.1 Basic quantities and the first two laws ... 112
   14.2 Constitutive equations and thermodynamic restrictions ... 114
   14.3 The heat equation ... 117

15	Thermodynamics of two-phase systems	118
	15.1 Basic quantities and the first two laws	118
	15.2 Local forms of the interfacial laws	122
16	Constitutive theory	124
	16.1 Constitutive equations for the bulk material	124
	16.2 The transition temperature	125
	16.3 Constitutive equations for the interface	126
17	Free-boundary problems	128
	17.1 Bulk equations and interface conditions	128
	17.2 Initial conditions and boundary conditions	129
	17.3 Free-boundary problems near the transition temperature for weak interfaces	130
	17.3.1 Approximate interface conditions	130
	17.3.2 Approximate free-boundary problems	131
	17.3.3 The first two laws for the approximate theories	132
	17.3.4 Growth theorems	134
	17.3.5 Perfect conductors	137
18	Instabilities induced by supercooling the liquid phase	138
	18.1 The one-dimensional problem: growth of the solid phase from a seed of zero measure	138
	18.2 Instability of a flat interface	140
References		142
Index		147

0
INTRODUCTION

In this treatise I discuss the dynamics of two-phase systems within the framework of modern continuum thermodynamics, assuming throughout a sharp[1] interface separating bulk phases. I consider only non-deformable continua, neglecting mass transport, and to avoid the geometrical complications associated with the motion of surfaces in $\mathbb{R}^3$, I restrict attention to evolution in the plane.[2] I consider two general theories: a purely mechanical theory that leads to a generalization of the classical curve-shortening equation

$$V = K \quad (V = \text{normal velocity}, \quad K = \text{curvature}) \tag{0.1}$$

of Mullins and Brakke;[3] and a theory of heat conduction that broadly generalized the classical Stefan theory.

In theories of continuum mechanics general principles common to large classes of materials are kept distinct from specific constitutive assumptions that differentiate between particular materials, and this is the point of view taken here. In particular, the basic ingredients of each of the theories developed here are:

(1) balance laws;

(2) an appropriate version of the second law of thermodynamics;

(3) constitutive equations.

The second law is used to ensure thermodynamically sound constitutive equations and these, with the balance laws, form the evolution equations of the theory.

Since the interface is an evolving curve in $\mathbb{R}^2$, its kinematics is fairly transparent. A detailed discussion of the kinematics is given in Part I.

Part II discusses the mechanical theory. The restriction to mechanics allows me to focus attention on the force system associated with interfacial motion. This system consists of capillary forces **C** that generalize the classical notion of surface tension, and interactive forces **B** that represent the interaction of the interface with the bulk material; **C** and **B** are vector-valued

---

[1] In the sense of Gibbs.
[2] Even so, many of the results apply almost without change to three-dimensional problems and to a multitude of two-phase phenomena including mass transfer by diffusion.
[3] Cf. Footnote 32.

functions related through the local force balance[4]

$$\mathbf{C}_s + \mathbf{B} = \mathbf{0}, \quad (s = \text{arc length}). \tag{0.2}$$

The interface is endowed with energy. This interfacial energy interacts with bulk energy through a mechanical version of the second law, the dissipation inequality, which asserts that the total energy of any control volume changes at a rate not greater than the power expended on it by interfacial forces. The dissipation inequality is used to develop suitable constitutive equations for the interface which, with the balance law (0.2), yields the following generalization of (0.1):

$$b(\theta)V = \{f(\theta) + f''(\theta)\}K - F. \tag{0.3}$$

Here θ is the angle to the interface normal, F is the difference in bulk energy between phases, $b(\theta) > 0$ is a material function that characterizes the kinetics, and the functional dependences of $b(\theta)$ and the interfacial energy $f(\theta)$ on θ allow for anisotropy.

When $f(\theta) + f''(\theta) > 0$ the evolution equation (0.3) is parabolic and fairly stable: it predicts the growth of sufficiently large crystals when F represents supercooling, and, as for the curve-shortening equation, it predicts a trend toward convexity for the interface. Material scientists use energies for which $f(\theta) + f''(\theta) < 0$ for certain intervals of θ; this renders the underlying equations *backward parabolic* and inherently unstable, but the unstable intervals may be excluded by inserting corners in the interface, and this leads to facets and (possibly infinitesimal) wrinklings. A central ingredient in the study of corners is the use of the integral form of (0.2) to ensure force balance across the corner. These issues are discussed in detail in Chapters 10–13. Chapter 14 develops a regularization of (0.3) that might provide an alternative method of treating the backward-parabolic regimes.

Part III presents a thermodynamical theory of interfacial evolution in the presence of bulk heat conduction. The interface is endowed with energy, entropy, and superficial force, but heat conduction within the interface is not included. The temperature is assumed to be continuous across the interface, but—to allow for phenomena such as supercooling and superheating—is otherwise unrestricted.

The theory is based on balance laws for force and energy in conjunction with an entropy-growth inequality that generalizes the classical Clausius–Duhem inequality. The chief results are a thermodynamically admissible constitutive theory for the interface, exact and approximate free-boundary

---

[4] Discussions of interfacial evolution are often based on the normal component of (0.2), derived as an Euler–Lagrange equation corresponding to the requirement that a global Gibbs function be stationary. This procedure has two chief deficiencies: it does not give the full balance law (0.1); and it is inapplicable in the presence of dissipation (cf. Footnote 20).

conditions at the interface, and a hierarchy of free-boundary problems at various levels of approximation. The simplest version of these problems—the Mullins–Sekerka problem—is essentially the classical Stefan problem with the free-boundary condition $u = 0$ for the temperature replaced by the condition $u = fK$, where f, a constant, is the interfacial free energy at $u = 0$.

I
KINEMATICS

1
CURVES

1.1 Preliminary definitions

A **curve** is a set $\imath$ in $\mathbb{R}^2$ together with a smooth map $p \mapsto \mathbf{r}(p)$ from an interval of $\mathbb{R}$ into $\mathbb{R}^2$ such that:

(1) $\imath$ is the range of $\mathbf{r}$;

(2)[5] $\mathbf{r}_p$ never vanishes;

(3) the domain of $\mathbf{r}$ is either all of $\mathbb{R}$ or a bounded interval $[P, Q]$;

(4) if the domain is $\mathbb{R}$, either $\mathbf{r}$ is periodic,[6] or $|\mathbf{r}(p)| \to \infty$ as $|p| \to \infty$.

The function $\mathbf{r}$ is a **parametrization** of $\imath$; the variable p is the **parameter**; the interval $[P, Q]$ (or $\mathbb{R}$) is the **parameter interval**; $\mathbf{r}(P)$ and $\mathbf{r}(Q)$ are the **initial** and **terminal points** (or collectively, the **endpoints**) of $\imath$, and we say that $\imath$ **is from** $\mathbf{r}(P)$ **to** $\mathbf{r}(Q)$.

We will generally refer to the set $\imath$ as the curve, the parametrization $\mathbf{r}$ being tacit. We will use the terms **bounded** or **unbounded** for $\imath$ in the sense of $\imath$ as a subset of $\mathbb{R}^2$, but we will refer to $\imath$ as **closed** if the domain of $\mathbf{r}$ is $\mathbb{R}$ and $\mathbf{r}$ is periodic. A curve $\imath$ has **endpoints** if the domain of $\mathbf{r}$ is a bounded interval; a non-closed curve $\imath$ is **simple** if $\mathbf{r}$ is one-to-one; a closed curve $\imath$ is **simple** if given any $p, q \in \mathbb{R}$, $\mathbf{r}(p) = \mathbf{r}(q)$ only when $p - q$ is a multiple of the minimal period of $\mathbf{r}$. Finally, $\imath$ is a **facet** if $\imath$ is a segment of a straight line.

Let $\imath$ be a curve. An **arc-length map** for $\imath$ is a smooth mapping $p \mapsto s(p)$ from the domain of $\mathbf{r}$ into $\mathbb{R}$ such that

$$s_p = |\mathbf{r}_p|. \tag{1.1}$$

We assume henceforth that an arc-length map is prescribed. Since $|\mathbf{r}_p| > 0$, the **arc length** $s = s(p)$ is an invertible function of p, and any function $\varphi(p)$ may be considered a function $\varphi(s)$, and vice versa.

The vector

$$\mathbf{T}(s) = \mathbf{r}_s(s) \tag{1.2}$$

defines a (unit) **tangent** to the curve in the direction of increasing p. We

---

[5] In general we use subscripts to denote *partial derivatives*; in this chapter we use subscripts also for total derivatives, since we want the resulting formulas to be applicable to evolving curves, which involve a second variable t.

[6] A function ϕ on $\mathbb{R}$ is **periodic** if there is a $\lambda > 0$ such that $\phi(p) = \phi(p + \lambda)$ for all $p \in \mathbb{R}$; λ is then a **period** of ϕ and the infimum of all periods is the **minimal period** of ϕ. (The minimal period of a curve $\mathbf{r}$ is strictly positive since $|\mathbf{r}_p| \neq 0$.)

CURVES

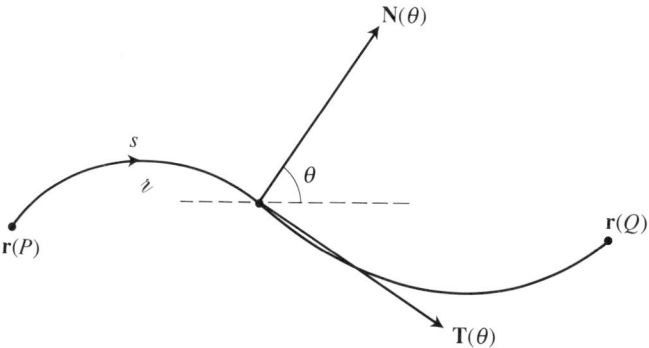

Fig. 1.1 Sign conventions for curves.

define a corresponding (unit) **normal** $\mathbf{N}(s)$ through the requirement that $\{\mathbf{T}, \mathbf{N}\}$ be a positively oriented orthonormal basis of $\mathbb{R}^2$, and define the **normal angle** $\theta(s)$ as a smooth function of s, through[7]

$$\mathbf{N} = (\cos\theta, \sin\theta), \qquad \mathbf{T} = (\sin\theta, -\cos\theta). \tag{1.3}$$

We will refer to the range of the function $s \mapsto \theta(s)$ as the **angle-set** (Figure 1.1). Note that $\mathbf{N}$ and $\mathbf{T}$ may be considered as functions of θ, in which case

$$\mathbf{N}_\theta = -\mathbf{T}, \qquad \mathbf{T}_\theta = \mathbf{N}. \tag{1.4}$$

The function

$$K(s) = \theta_s(s) \tag{1.5}$$

is the **curvature**; by (1.4), $K(s)$ obeys the Frenet formulas:

$$\mathbf{N}_s = -K\mathbf{T}, \qquad \mathbf{T}_s = K\mathbf{N}. \tag{1.6}$$

Let $\imath$ be simple and either closed or unbounded. Then, by the Jordan curve theorem, $\imath$ is a **boundary curve**; that is, $\imath$ divides $\mathbb{R}^2$ into two regions. In particular, $\imath$ is the boundary of a region Ω whose outward normal coincides with the normal to $\imath$. We will refer to Ω as the **reference region** for $\imath$.

1.2 Convex curves

A curve is **convex** if its curvature never vanishes. By (1.5), for such a curve the mapping $s \mapsto \theta(s)$ is invertible, and we may use θ in place of s or p as independent variable. In particular, we may parametrize the curve itself by θ, giving a function $\mathbf{r}(\theta)$; granted this, (1.2) and (1.5) yield

$$\mathbf{r}_\theta = K^{-1}\mathbf{T}. \tag{1.7}$$

[7] This defines $\theta(s)$ (the angle from $(1, 0) \in \mathbb{R}^2$ to $\mathbf{N}(s)$) up to a multiple of 2π.

Strictly speaking, the function $\theta \mapsto \mathbf{r}(\theta)$ is a parametrization of $\imath$ only if $K > 0$, for otherwise θ does not increase with increasing s; even so, we will refer to $\mathbf{r}(\theta)$ as the **angle-parametrization** of $\imath$.

We write

$$\varphi_{\text{norm}} = \boldsymbol{\varphi} \cdot \mathbf{N}, \qquad \varphi_{\text{tan}} = \boldsymbol{\varphi} \cdot \mathbf{T} \tag{1.8}$$

for the normal and tangential components of a vector field $\boldsymbol{\varphi}$. In particular,

$$\mathrm{p} = r_{\text{norm}} \tag{1.9}$$

is the **support function** of the curve, a field that is useful in the study of convex curves, for then, by (1.4) and (1.7),

$$\mathbf{r} = \mathrm{p}\mathbf{N} - \mathrm{p}_\theta \mathbf{T}, \qquad \mathrm{p}_{\theta\theta} + \mathrm{p} = -K^{-1}. \tag{1.10}$$

We now give three useful lemmas concerning convex curves.

(1A) Lemma
(a) *Two convex curves with the same angle-set and the same curvature are equal modulo a translation.*

(b) *if $\imath$ is a convex curve with curvature $K(\theta)$, and if $\mathrm{p}(\theta)$ is any solution of $(1.10)_2$ on the angle-set of $\imath$, then $(1.10)_1$ is the angle-parametrization of a curve that differs from $\imath$ by at most a translation.*

PROOF. (a) The general solution of $\mathrm{p}_{\theta\theta} + \mathrm{p} = 0$ is $\mathrm{p}(\theta) = \mathbf{a} \cdot \mathbf{N}(\theta)$ with $\mathbf{a}$ an arbitrary vector. By $(1.10)_2$, the difference between the support functions of the two curves must have the form $\mathbf{a} \cdot \mathbf{N}(\theta)$, and this with $(1.10)_1$ implies that $\mathbf{r}_1(\theta) - \mathbf{r}_2(\theta) = \mathbf{a}$. (b) The proof is the same as that for (a) except we use the difference between p and the support function for $\imath$. ∎

(1B) Lemma *Consider a curve whose curvature is not identically zero, and suppose that the curvature has the form $K(\theta(s))$ with $K(\theta)$ a smooth function on the angle-set. Then the curve is convex.*

PROOF. Let Y denote the angle-set, and let Γ be a connected component of the set $\{\theta \in Y : K(\theta) \neq 0\}$. We must show that $\Gamma = Y$. Assume that $\Gamma \neq Y$. Then there is a boundary point θ_0 of Γ in $\mathbb{R}$ with $\theta_0 \in Y$ and $K(\theta_0) = 0$. Since $K(\theta)$ is smooth up to θ_0, $|K(\theta)| \leq C|\theta - \theta_0|$ near θ_0. But on Γ, $s = s(\theta)$ and $s'(\theta) = K(\theta)^{-1}$, so that $|s(\theta)| \to \infty$ as $\theta \to \theta_0$, a contradiction. ∎

A curve $\imath$ is of **class R** if $\imath$ is convex and either $\imath$ has endpoints or $\imath$ is a boundary curve.

(1C) Lemma *Let $K(\theta)$ be a smooth function on the closure of an interval $\mathcal{O}$ of $\mathbb{R}$. Then $K(\theta)$ is the curvature and $\mathcal{O}$ the angle-set of a curve $\imath$ of class R if and only if $K(\theta)$ does not vanish on $\mathcal{O}$ and one of (a), (b), or (c) is satisfied:*

(a) *$\mathcal{O}$ is a bounded interval $[\theta_1, \theta_2]$;*

(b) *$\mathcal{O} = \mathbb{R}$ and $K(\theta)$ is 2π-periodic with*

$$\int_0^{2\pi} K(\theta)^{-1} e^{i\theta} \, d\theta = 0; \tag{1.11}$$

(c) *$\mathcal{O}$ is a bounded interval (θ_1, θ_2) with $\theta_2 - \theta_1 \leq \pi$ and*

$$K(\theta_1 + 0) = K(\theta_2 - 0) = 0.$$

In case (a), $\imath$ is a convex curve with endpoints; in case (b), $\imath$ is a convex, closed boundary curve; in case (c), $\imath$ is a convex, unbounded boundary curve.

PROOF. Note first that if $g(\theta)$ is a smooth 2π-periodic function on $\mathbb{R}$, then

$$\int_0^{2\pi} [g(\theta) + g''(\theta)] e^{i\theta} \, d\theta = 0, \tag{1.12}$$

an assertion that follows upon integrating the term $g''(\theta) e^{i\theta}$ twice by parts. (The prime denotes differentiation with respect to θ.)

Suppose that $K(\theta)$ is the curvature of a class R curve $\imath$ with angle-set $\mathcal{O}$. Then, by definition, either $\imath$ has endpoints, or $\imath$ is a closed boundary curve, or $\imath$ is an unbounded boundary curve. If $\imath$ has endpoints, then $\mathcal{O}$ is necessarily a bounded interval $[\theta_1, \theta_2]$; hence (a) is satisfied. If $\imath$ is a closed boundary curve, then $\mathcal{O} = \mathbb{R}$ and both $K(\theta)$ and $\mathbf{r}(\theta)$ are 2π-periodic, as is the support function $p(\theta)$. Thus (1.11) is a consequence of (1.12) and (1.10)$_2$, and (b) is satisfied. Finally, if $\imath$ is an unbounded boundary curve, then $|\mathbf{r}(p)| \to \infty$ as $p \to \pm\infty$ and

$$|s(p)| \to \infty \quad \text{as} \quad p \to \pm\infty. \tag{1.13}$$

Further, since $\imath$ is simple, $\mathcal{O}$ is a bounded interval (θ_1, θ_2) with $\theta_2 - \theta_1 \leq \pi$. Moreover, $s'(\theta) = K(\theta)^{-1}$ and, by (1.13), $|s(\theta)| \to \infty$ as $\theta \to \theta_1, \theta_2$; thus, since $K(\theta)$ is smooth up to θ_1 and θ_2, $K(\theta_1 + 0) = K(\theta_2 - 0) = 0$. Therefore (c) is satisfied.

Conversely, suppose that $K(\theta)$ is non-vanishing on $\mathcal{O}$ and that either (a), (b), or (c) is satisfied. Let $p(\theta)$ be any solution of (1.10)$_2$ on $\mathcal{O}$ and define $\mathbf{r}(\theta)$ by (1.10)$_1$, so that $\mathbf{r}(\theta)$ satisfies (1.7). Let $s = s(\theta)$ be any solution of $s'(\theta) = K(\theta)^{-1}$ on $\mathcal{O}$. Since $K(\theta)$ is smooth and non-vanishing on $\mathcal{O}$, $s = s(\theta)$ is an invertible map with inverse $\theta = \theta(s)$ consistent with

$$\mathbf{r}_s = \mathbf{T}, \quad \theta_s = K.$$

Let $\imath$ denote the range of $\mathbf{r}(\theta)$, and let $\jmath$ denote the range of $s(\theta)$.

Assume that (a) is satisfied. Then $\mathscr{S}$ is a compact interval, so that $\imath$ is a convex curve with $\mathbf{r}(s)$ an arc-length parametrization; hence $\imath$ is of class R.

Assume that (b) is satisfied. Then $(1.3)_2$, (1.7), and (1.11) imply that $\mathbf{r}(\theta)$ is 2π-periodic, and since $K(\theta)$ is also 2π-periodic, we may conclude that $\mathbf{r}(s)$ is periodic. Thus $\imath$ is closed. Further, since $\imath$ is convex and $\mathbf{r}(0) = \mathbf{r}(2\pi)$, $\imath$ is simple. Thus $\imath$ is a convex, closed boundary curve.

Assume that (c) is satisfied. Since $K(\theta)$ is smooth up to θ_1 and θ_2, $|K(\theta)| \leq C|\theta - \theta_0|$ for all θ near θ_1 and θ_2. Thus, since $s_\theta = K^{-1}$, $\mathbf{r}_\theta = K^{-1}\mathbf{T}$, and $K(\theta_1) = K(\theta_2) = 0$, it follows that $|s(\theta)| \to \infty$ and $|\mathbf{r}(\theta)| \to \infty$ as $\theta \to \theta_1$, θ_2. Thus $\imath$ is a convex, unbounded curve. Since $\theta_2 - \theta_1 \leq \pi$, $\imath$ is simple. Thus $\imath$ is a boundary curve. ∎

1.3 Integrals

Let $\imath$ be a curve with arc length lying in an interval $[S_1, S_2]$. Given a function $\varphi(s)$, we use the notation[8]

$$\int_\imath \varphi \, ds = \int_{S_1}^{S_2} \varphi(s) \, ds,$$

$$\int_{\partial \imath} \varphi = \varphi(S_2) - \varphi(S_1). \tag{1.14}$$

Then

$$\int_{\partial \imath} \varphi = \int_\imath \varphi_s \, ds. \tag{1.15}$$

(1D) Area identity Let $\imath$ be a boundary curve whose reference region Ω is bounded. Then

$$\text{area}(\Omega) = \frac{1}{2} \int_\imath \mathbf{r} \cdot \mathbf{N} \, ds. \tag{1.16}$$

PROOF. Let $\mathbf{z}(\mathbf{x}) = \mathbf{x}$ for all $\mathbf{x} \in \mathbb{R}^2$. Then, using the divergence theorem,

$$2 \, \text{area}(\Omega) = \int_\Omega \text{div } \mathbf{z} \, da = \int_\imath \mathbf{r} \cdot \mathbf{N} \, ds. \quad \blacksquare$$

1.4 Piecewise-smooth curves

We now extend some of the previous definitions and results to curves that are continuous but not smooth. For convenience, we write 'PS' as an abbreviation for '**piecewise smooth**'.

---

[8] This definition of the 'integral over $\partial \imath$' makes the statement of integral balance laws in $\mathbb{R}^2$ suggestive of their three-dimensional counterparts.

CURVES

Let $\imath = \{\imath_1, \imath_2, \ldots\}$ denote a finite or countably infinite list of curves $\imath_i$, called **arcs** of $\imath$, with $\mathbf{r}_i$ the parametrization and $[P_i, Q_i]$ the parameter interval for $\imath_i$. Then $\imath$ is a **PS curve** if:

(1) at each **juncture** i (that is, each i with $\imath_i$ and $\imath_{i+1}$ arcs of $\mathbf{r}$),

$$\mathbf{r}_i(Q_i) = \mathbf{r}_{i+1}(P_{i+1});$$

(2) there is an integer N such that either

(a) $\imath$ consists of N arcs, in which case $\imath_1$ and $\imath_N$ are **terminal arcs**, and $\imath$ has endpoints $\mathbf{r}_{\text{init}} = \mathbf{r}(P_1)$ and $\mathbf{r}_{\text{term}} = \mathbf{r}(Q_N)$; or

(b) $\imath$ consists of an infinite number of arcs, but $\mathbf{r}_i = \mathbf{r}_{N+i}$ for all i; in this case $\imath$ is **closed**, the smallest integer N is the **essential number of arcs** of $\imath$, and the junctures $i = 1, 2, \ldots, N$ are the **essential junctures**.

In addition: $\imath_i$ is an **internal arc** if both i and $i - 1$ are junctures.

We define a **PS boundary curve** to be a PS curve that forms the boundary of a region in $\mathbb{R}^2$. Most of the definitions concerning (smooth) curves extend in a natural manner to PS curves; we will not repeat these.

By a fully **faceted curve** we mean a PS curve $\imath = \{\imath_1, \imath_2, \ldots\}$ each of whose arcs is a *facet*. An example of a fully faceted curve is a **wrinkled curve between the normal angles α and β**, or, more succinctly, an α, β **wrinkled curve**: here there are constant angles α and β such that

$$\theta(s) = \alpha \quad \text{on } \imath_i \text{ for } i \text{ odd,}$$
$$\theta(s) = \beta \quad \text{on } \imath_i \text{ for } i \text{ even;}$$

the facets with normal angle α are then called α-**facets**, and analogously for β.

Let ω be an α, β wrinkled curve, $0 < \beta - \alpha < \pi$, with $\mathbf{x}$ as the initial point and $\mathbf{y}$ as the terminal point. Then the **total lengths** $L_\alpha(\omega)$ and $L_\beta(\omega)$ of α-facets and β-facets depend on ω only through the distance L and the normal angle θ_0 of the oriented straight line ℓ from $\mathbf{x}$ to $\mathbf{y}$:

$$L\mathbf{T}(\theta_0) = L_\alpha(\omega)\mathbf{T}(\alpha) + L_\beta(\omega)\mathbf{T}(\beta), \tag{1.17}$$

and, trivially, for any function $\varphi(\theta)$,

$$\int_\omega \varphi(\theta)\, ds = \varphi(\alpha)L_\alpha(\omega) + \varphi(\beta)L_\beta(\omega). \tag{1.18}$$

Let $\mu_\alpha(\theta)$ and $\mu_\beta(\theta)$ be defined by

$$\mathbf{T}(\theta) = \mu_\alpha(\theta)\mathbf{T}(\alpha) + \mu_\beta(\theta)\mathbf{T}(\beta) \tag{1.19}$$

for any angle $\theta \in [\alpha, \beta]$; then

$$\mu_\alpha(\theta_0) = \frac{L_\alpha(\omega)}{L}, \qquad \mu_\beta(\theta_0) = \frac{L_\beta(\omega)}{L}, \tag{1.20}$$

so that $\mu_\alpha(\theta_0)$ and $\mu_\beta(\theta_0)$ represent the total lengths of α-facets and β-facets of w per unit length of ℓ. These numbers depend on the α, β wrinkled curve w only through the normal angle θ_0 of ℓ; we will refer to $\mu_\alpha(\theta_0)$ and $\mu_\beta(\theta_0)$ (defined by (1.19)) as the **facet-densities** of α-facets and β-facets corresponding to θ_0. Note that if we add (1.19) to itself differentiated twice with respect to θ, we conclude, with the aid of (1.4), that

$$\mu_\alpha(\theta) + \mu_\alpha''(\theta) = 0, \qquad \mu_\beta(\theta) + \mu_\beta''(\theta) = 0. \tag{1.21}$$

1.5 Infinitesimally wrinkled curves[9]

It is often convenient to view a PS curve $\imath$ as an α, β wrinkled curve $\imath_{\alpha\beta}$ ($0 < \beta - \alpha < \pi$) whose facets are *infinitesimally small*. Proceeding formally, consider an infinitesimal piece $\mathrm{d}\imath_{\alpha\beta}$ of $\imath_{\alpha\beta}$ at s, let $\mathrm{d}L_\alpha$ and $\mathrm{d}L_\beta$ denote the total lengths of facets of normal angle α and normal angle β for $\mathrm{d}\imath_{\alpha\beta}$, and let $\mathrm{d}L$ denote the length of $\mathrm{d}\imath$. Then, by (1.17),

$$\mathbf{T}(\theta(s))\,\mathrm{d}L = \mathbf{T}(\alpha)\,\mathrm{d}L_\alpha + \mathbf{T}(\beta)\,\mathrm{d}L_\beta,$$

with $\theta(s)$ the normal angle of $\imath$ at s. Thus $\theta(s)$ must belong to the interval $[\alpha, \beta]$, and

$$\frac{\mathrm{d}L_\alpha}{\mathrm{d}L} = \mu_\alpha(\theta(s)), \qquad \frac{\mathrm{d}L_\beta}{\mathrm{d}L} = \mu_\beta(\theta(s)), \tag{1.22}$$

with $\mu_\alpha(\theta)$ and $\mu_\beta(\theta)$ defined by (1.19); $\mu_\alpha(\theta(s))$ and $\mu_\beta(\theta(s))$ therefore represent the densities of α-facets and β-facets on $\imath_{\alpha\beta}$ at the arc length s.

Precisely, an **infinitesimally wrinkled curve** $\imath_{\alpha\beta}$ is a PS curve $\imath$ together with angles α and β, $0 < \beta - \alpha < \pi$, such that

$$\text{the angle-set of } \imath \text{ is contained in } [\alpha, \beta]. \tag{1.23}$$

Guided by (1.22), we define the total lengths of α-facets and β-facets of $\imath_{\alpha\beta}$ through

$$L_\alpha(\imath_{\alpha\beta}) = \int_\imath \mu_\alpha(\theta)\,\mathrm{d}s, \qquad L_\beta(\imath_{\alpha\beta}) = \int_\imath \mu_\beta(\theta)\,\mathrm{d}s, \tag{1.24}$$

and, for any function $\varphi(\theta)$, we write

$$\int_{\imath_{\alpha\beta}} \varphi(\theta)\,\mathrm{d}s = L_\alpha(\imath_{\alpha\beta})\varphi(\alpha) + L_\beta(\imath_{\alpha\beta})\varphi(\beta)$$

$$= \int_{\imath_{\alpha\beta}} \{\varphi(\alpha)\mu_\alpha(\theta) + \varphi(\beta)\mu_\beta(\theta)\}\,\mathrm{d}s. \tag{1.25}$$

---

[9] *Generalized curves* in the sense of Young (1969); *varifolds* in the sense of Almgren (1966).

The next proposition shows that—with respect to integration—infinitesimal wrinkled curves and wrinkled curves exhibit similar behaviour.

(1E) Proposition *Let $\imath_{\alpha\beta}$ be an infinitesimal wrinkled curve and let ω be an α, β wrinkled curve whose initial and terminal points coincide with those of $\imath_{\alpha\beta}$. Then*

$$L_\alpha(\imath_{\alpha\beta}) = L_\alpha(\omega), \qquad L_\beta(\imath_{\alpha\beta}) = L_\beta(\omega), \qquad (1.26)$$

and, for any function $\varphi(\theta)$,

$$\int_{\imath_{\alpha\beta}} \varphi(\theta)\,ds = \int_\omega \varphi(\theta)\,ds. \qquad (1.27)$$

PROOF. We have only to establish (1.26), since (1.18), (1.25), and (1.26) imply (1.27). Let L denote the length and θ_0 the normal angle of the oriented straight line whose initial and terminal points coincide with those of $\imath_{\alpha\beta}$ (and hence ω). Then, since

$$\int_\imath \mathbf{T}(\theta(s))\,ds = L\mathbf{T}(\theta_0),$$

(1.17), (1.19), and (1.24) yield (1.26). ∎

2
EVOLVING CURVES

This chapter discusses the kinematics of curves that evolve smoothly in time, and forms the basis of our discussion of phase interfaces.

2.1 Definitions

Let t denote the time. An **evolving curve** $\imath$ is a family $\imath(t)$, $0 \leq t < T$, of *curves* $\imath(t)$ together with a smooth map $(p, t) \mapsto \mathbf{r}(p, t)$ such that:

(1) $\mathbf{r}(\cdot, t)$ is a parametrization of $\imath(t)$ at each t;
(2) the domain of $\mathbf{r}$ is either $\mathbb{R} \times [0, T)$ or a set of the form

$$\{(p, t): p \in [P(t), Q(t)], t \in [0, T)\},$$

where $P, Q: [0, T) \to \mathbb{R}$ ($P < Q$) are smooth functions.

The time T is the **duration** of $\imath$.

In certain discussions (Chapter 7) involving the calculus of variations the underlying 'time interval' will be $(-T, T)$ rather than $[0, T)$; this difference leads to no essential changes in the results of the present chapter.

An evolving curve $\imath$ is an **evolving interface** if $\imath(t)$ is a boundary curve at each t; the *reference region* $\Omega(t)$ for $\imath(t)$ then has $\imath(t)$ as its boundary curve, with the outward normal to $\Omega(t)$ the normal to $\imath(t)$.

Let $\imath$ be an evolving curve: $\imath$ is bounded, unbounded, closed, simple, convex, or has endpoints, according as $\imath(t)$ has that property for each $t \in [0, T)$ (for $\imath$ closed we add the requirement that the period be a smooth function of time); a *bounded* evolving curve $\imath_0$ is an *evolving subcurve* of $\imath$ if, modulo a translation of time, the parametrization $\mathbf{r}_0$ of $\imath_0$ is a restriction of the parametrization of $\imath$.

Let an evolving curve $\imath$ be given. An **arc-length map** for $\imath$ is a smooth mapping $s(p, t)$ such that $s(\cdot, t)$ is an arc-length map for the curve $\imath(t)$ at each t. It is not difficult to construct an arc-length map for $\imath$, and any two such maps differ by a smooth function of time. We assume henceforth that an arc-length map is prescribed. Since $s = s(p, t)$ is an invertible function of p, any function $\varphi(p, t)$ may be considered a function $\varphi(s, t)$, and vice versa. We will refer to $\varphi(s, t)$ as the **arc-length description** of φ.

2.2 Transport identities

Let $\imath$ be an evolving curve with parametrization $\mathbf{r}(s, t)$ expressed in terms of arc length s, and let $\mathbf{T}(s, t)$ and $\mathbf{N}(s, t)$ denote the corresponding tangent and normal. We refer to

$$V(s, t) = \mathbf{r}_t(s, t) \cdot \mathbf{N}(s, t), \tag{2.1}$$

as the **normal velocity** and to

$$v(s, t) = -\mathbf{r}_t(s, t) \cdot \mathbf{T}(s, t), \tag{2.2}$$

as the **arc velocity**, where $\mathbf{r}_t$ is the time derivative holding s fixed.

Let $\mathscr{S}$ denote the domain of $\mathbf{r}(s, t)$. Choose $(s_0, t_0) \in \mathscr{S}$ and let $S(t)$ be defined in a neighbourhood of t_0 and satisfy $S(t_0) = s_0$. Then $S(t)$ is the **normal arc-length trajectory** through s_0 at time t_0 if

$$\mathbf{T}(S(t), t) \cdot \frac{d\mathbf{r}(S(t), t)}{dt} = 0, \tag{2.3}$$

or equivalently (by (2.2)) if

$$S^{\bullet}(t) = v(S(t), t), \qquad S(t_0) = s_0. \tag{2.4}$$

(Here and in what follows we write $\phi^{\bullet}(t)$ for the derivative of a function $\phi(t)$ of time alone.)

Let φ be a smooth function on $\mathscr{S}$. Then the **normal time derivative** $\varphi^{\circ}(s, t)$ of φ at (s, t) is defined as follows:

$$\varphi^{\circ}(s, t) = \left. \frac{d\varphi(S(\tau), \tau)}{d\tau} \right|_{\tau = t},$$

where $S(\tau)$ is the normal arc-length trajectory through (s, t). Then

$$\varphi^{\circ} = \varphi_t + v\varphi_s, \tag{2.5}$$

with φ_t the time derivative holding s fixed.

By (1.2), (2.1)–(2.3), and (2.5),

$$\mathbf{r}^{\circ} \cdot \mathbf{N} = \mathbf{r}_t \cdot \mathbf{N} = V, \qquad \mathbf{r}^{\circ} \cdot \mathbf{T} = 0; \tag{2.6}$$

hence

$$\mathbf{r}^{\circ}(p, t) = V(p, t)\mathbf{N}(p, t). \tag{2.7}$$

Note also that, by (2.5),

$$(\varphi^{\circ})_s = (\varphi_t + v\varphi_s)_s = \varphi_{st} + v\varphi_{ss} + v_s\varphi_s = (\varphi_s)^{\circ} + v_s\varphi_s. \tag{2.8}$$

(2A) Transport identities

$$v_s = KV, \qquad \theta^{\circ} = V_s, \qquad K^{\circ} = V_{ss} + K^2 V. \tag{2.9}$$

PROOF. Since $\mathbf{r}_{ts} = \mathbf{r}_{st} = \mathbf{T}_t$ is perpendicular to $\mathbf{T}$, (1.6), (2.2), and (2.6) yield $(2.9)_1$.

Let $\mathbf{e}$ be a fixed unit vector. Then, by (1.2), (1.3), and (2.8),

$$(\mathbf{r}_s)^\circ \cdot \mathbf{e} = (\mathbf{T} \cdot \mathbf{e})^\circ = (\mathbf{N} \cdot \mathbf{e})\theta^\circ,$$

$$(\mathbf{r}_s)^\circ \cdot \mathbf{e} = (\mathbf{r}^\circ)_s \cdot \mathbf{e} - v_s \mathbf{T} \cdot \mathbf{e},$$

while (1.6), (2.7), and $(2.9)_1$ imply

$$(\mathbf{r}^\circ)_s = V_s \mathbf{N} + v_s \mathbf{T}.$$

The last three relations yield $(2.9)_2$, since $\mathbf{e}$ is arbitrary (cf. (2.5)). Finally, $(2.9)_3$ follows from (2.8) with $\varphi = \theta$, (1.5), $(2.9)_1$, and $(2.9)_2$. ∎

Note that for an evolving facet (cf. (1.5) and $(2.9)_1$),

$$K = \theta_s = 0, \qquad v_s = 0. \tag{2.10}$$

Consider now an evolving curve $\imath$ with endpoints, and let $[S_1(t), S_2(t)]$ denote the interval of arc lengths for $\imath$ at time t. Further, let $\mathbf{R}_1(t)$ and $\mathbf{R}_2(t)$ denote the initial and terminal points of $\imath$ with $\mathbf{T}_1(t)$, $\mathbf{T}_2(t)$, $\mathbf{N}_1(t)$, and $\mathbf{N}_2(t)$ the corresponding tangents and normals. The motion of the endpoints is then characterized by the **endpoint velocities**

$$\mathbf{v}_{\partial \imath}(s, t) = \mathbf{R}_i^\bullet(t), \qquad s = S_i(t), \tag{2.11}$$

and the **tangential endpoint velocities**

$$v_{\partial \imath (\tan)}(s, t) = \mathbf{T}_i(t) \cdot \mathbf{R}_i^\bullet(t), \qquad s = S_i(t), \tag{2.12}$$

$i = 1, 2$. Since $\mathbf{R}_i(t) = \mathbf{r}(S_i(t), t)$,

$$\mathbf{R}_i^\bullet(t) = \mathbf{r}_s(S_i(t), t) S_i^\bullet(t) + \mathbf{r}_t(S_i(t), t), \tag{2.13}$$

and, since $\mathbf{r}_s$ is tangential, (2.1) yields

$$V(s, t) = \mathbf{N}_i(t) \cdot \mathbf{R}_i^\bullet(t) \tag{2.14}$$

at $s = S_i(t)$. Further, (2.2), (2.12), and (2.13) imply that

$$v_{\partial \imath (\tan)}(s, t) = S_i^\bullet(t) - v(s, t). \tag{2.15}$$

(2B) Proposition *Consider the endpoint at $s = S_i(t)$. Let $\Theta_i(t)$ denote the normal angle of the endpoint. Then, at $s = S_i(t)$,*

$$\mathbf{R}_i^\bullet(t) = V(s, t) \mathbf{N}_i(t) + v_{\partial \imath (\tan)}(s, t) \mathbf{T}_i(t),$$
$$\Theta_i^\bullet(t) = V_s(s, t) + v_{\partial \imath (\tan)}(s, t) K(s, t). \tag{2.16}$$

PROOF. The first of (2.16) follows from (2.12), (2.13), and (2.14). To derive the second we differentiate $\Theta_i(t) = \theta(S_i(t), t)$ and use (2.5), $(2.9)_2$, and (2.15). ∎

The endpoint $\mathbf{R}_i(t)$ is a **normal trajectory** if $\mathbf{R}_i^{\cdot}(t) = V(S_i(t), t)\mathbf{N}_i(t)$. In view of the relations derived above, this condition is equivalent to each of the following four conditions:

$$v_{\partial \imath(\tan)}(S_i(t), t) = 0, \qquad S_i^{\cdot}(t) = v(s, t),$$
$$\Theta_i^{\cdot}(t) = V_s(s, t), \qquad P_i^{\cdot}(t) = 0. \tag{2.17}$$

Note that, by $(2.10)_2$ and (2.15), for $\imath$ an evolving facet, the length $L(t) = S_2(t) - S_1(t)$ satisfies

$$L^{\cdot}(t) = v_{\partial \imath(\tan)}(S_2(t), t) - v_{\partial \imath(\tan)}(S_1(t), t). \tag{2.18}$$

For a convex evolving curve the mapping $s \mapsto \theta(s, t)$ is invertible and we may use θ and t in place of s and t as independent variables. Then

$$K^{\circ} = K_t + K_{\theta}\theta^{\circ} \tag{2.19}$$

(with K_t the derivative of K with respect to t holding θ fixed).

(2C) Proposition *For a convex evolving curve with curvature, normal velocity, and arc velocity expressed as functions of (θ, t),*

$$K_t = K^2(V_{\theta\theta} + V), \qquad v_{\theta} = -V. \tag{2.20}$$

In addition, if the curve has endpoints, and if the normal angle θ at an endpoint $\mathbf{R}(t)$ has the constant value θ_0, then

$$\mathbf{R}^{\cdot}(t) = V(\theta_0, t)\mathbf{N}(\theta_0) - V_{\theta}(\theta_0, t)\mathbf{T}(\theta_0). \tag{2.21}$$

PROOF. Clearly,

$$V_s = V_{\theta}K, \qquad V_{ss} = V_{\theta\theta}K^2 + V_{\theta}K_{\theta}K,$$

and these relations, (2.19), and $(2.9)_{2,3}$ yield $(2.20)_1$. On the other hand, $v_s = v_{\theta}K$ and $(2.20)_2$ follows from $(2.9)_1$. Finally, (2.16) with $\Theta^{\cdot} = 0$ and $V_s = V_{\theta}K$ imply (2.21). ∎

2.3 Integral identities

Let $\imath$ be an evolving curve with endpoints, and let $[P_1(t), P_2(t)]$ and $[S_1(t), S_2(t)]$ denote the corresponding parameter and arc-length intervals. Then, using the definitions (1.14), (2.11), and (2.12),

$$\int_{\partial \imath(t)} \boldsymbol{\varphi} \cdot \mathbf{v}_{\partial \imath} = (\boldsymbol{\varphi} \cdot \mathbf{v}_{\partial \imath})(S_2(t), t) - (\boldsymbol{\varphi} \cdot \mathbf{v}_{\partial \imath})(S_1(t), t)$$
$$= \boldsymbol{\varphi}(S_2(t), t) \cdot \mathbf{R}_2^{\cdot}(t) - \boldsymbol{\varphi}(S_1(t), t) \cdot \mathbf{R}_1^{\cdot}(t), \tag{2.22a}$$

$$\int_{\partial\imath(t)} \varphi \mathbf{v}_{\partial\imath(\tan)} = (\varphi \mathbf{v}_{\partial\imath(\tan)})(S_2(t), t) - (\varphi \mathbf{v}_{\partial\imath(\tan)})(S_1(t), t),$$

$$= \varphi(S_2(t), t)\mathbf{T}_2(t) \cdot \mathbf{R}_2^{\boldsymbol{\cdot}}(t) - \varphi(S_1(t), t)\mathbf{T}_1(t) \cdot \mathbf{R}_1^{\boldsymbol{\cdot}}(t). \quad (2.22\text{b})$$

The 'integral' involving φ represents the rate at which φ is *carried into* $\imath(t)$ across $\partial\imath(t)$ due to the tangential motion of $\partial\imath(t)$; this integral vanishes when the endpoints are normal trajectories. Also, by (1.8) and (2.16)$_1$,

$$\int_{\partial\imath(t)} \boldsymbol{\varphi} \cdot \mathbf{v}_{\partial\imath} = \int_{\partial\imath(t)} [\varphi_{\tan}\mathbf{v}_{\partial\imath(\tan)} + \varphi_{\text{norm}} V]. \quad (2.23)$$

(2D) Transport theorem for integrals

$$\frac{d}{dt}\int_{\imath(t)} \varphi \, ds = \int_{\imath(t)} (\varphi^\circ - \varphi K V) \, ds + \int_{\partial\imath(t)} \varphi \mathbf{v}_{\partial\imath(\tan)}. \quad (2.24)$$

PROOF. By (2.5) the left side of (2.24) is given by (suppressing the argument t where convenient)

$$\frac{d}{dt}\int_{S_1(t)}^{S_2(t)} \varphi(s, t) \, ds = S_2^{\boldsymbol{\cdot}}\varphi(S_2) - S_1^{\boldsymbol{\cdot}}\varphi(S_1) + \int_{S_1}^{S_2} (\varphi^\circ - v\varphi_s) \, ds.$$

If we integrate the term involving $v\varphi_s$ by parts and appeal to (2.9)$_1$ and (2.15), we arrive at (2.24). ∎

If we take $\varphi = 1$ in (2.24), we arrive at a transport identity for the length $L(t)$ of $\imath(t)$:

$$L^{\boldsymbol{\cdot}}(t) = -\int_{\imath(t)} KV \, ds + \int_{\partial\imath(t)} \mathbf{v}_{\partial\imath(\tan)}. \quad (2.25)$$

(2E) Transport theorem for area and perimeter Let $\imath$ be a simple, closed evolving curve, let $\Omega(t)$ be the bounded region enclosed by $\imath$, and assume that the normal $\mathbf{N}$ to $\imath$ is the outward normal to $\partial\Omega(t)$. Then the area $A(t)$ of $\Omega(t)$ and the perimeter $L(t)$ of $\partial\Omega(t)$ evolve according to

$$A^{\boldsymbol{\cdot}}(t) = \int_{\imath(t)} V \, ds.$$
$$L^{\boldsymbol{\cdot}}(t) = -\int_{\imath(t)} KV \, ds, \quad (2.26)$$

PROOF. The identity (2.26)$_2$ follows fro (2.25). To prove (2.26)$_1$, note first that, by (1.4) and (2.9),

$$\mathbf{N}^\circ = -\mathbf{T}V_s. \tag{2.27}$$

Thus, using (1.16) and (2.24) in conjunction with (2.7),

$$2A^\cdot(t) = \int_{\imath(t)} (\mathbf{r}^\circ \cdot \mathbf{N} + r \cdot \mathbf{N}^\circ - KV\mathbf{r}\cdot\mathbf{N})\,ds = \int_{\imath(t)} (V - \mathbf{r}\cdot\mathbf{T}V_s - KV\mathbf{r}\cdot\mathbf{N})\,ds.$$

On the other hand, by (1.2) and (1.6),

$$-\int_{\imath(t)} \mathbf{r}\cdot\mathbf{T}V_s\,ds = \int_{\imath(t)} (\mathbf{r}\cdot\mathbf{T})_s V\,ds = \int_{\imath(t)} (V + KV\mathbf{r}\cdot\mathbf{N})\,ds,$$

and $(2.26)_1$ follows. ∎

By (2.25), the transport identity $(2.26)_2$ is also satisfied by an evolving curve whose endpoints are normal trajectories; this yields the formal identity: $(ds)^\circ = -KV\,ds$.

2.4 Steadily evolving interfaces

By a **steadily evolving interface** we mean an *evolving interface* $\imath$ with arc-length parametrization of the form:

$$\mathbf{r}(s,t) = \mathbf{r}_0(s) + t\mathbf{U}, \qquad \mathbf{U} \neq 0; \tag{2.28}$$

the vector $\mathbf{U}$ is then the **steady velocity**; the curve $\imath_0$ parametrized by $\mathbf{r}_0$ is the **portrait**.

(2F) Proposition *For a steadily evolving interface the normal velocity $V(s)$, the curvature, $K(s)$, the normal $\mathbf{N}(s)$, the tangent $\mathbf{T}(s)$, and the normal angle $\theta(s)$ are independent of time, and*

$$V(s) = \mathbf{U}\cdot\mathbf{N}(s). \tag{2.29}$$

PROOF. By (2.28), $\mathbf{T}(s)$ and (hence) $\mathbf{N}(s)$ and $\theta(s)$ are independent of t. Further, since $V\mathbf{N} = \mathbf{r}^\circ = \mathbf{r}_t + v\mathbf{r}_s$ and $\mathbf{T} = \mathbf{r}_s$, $V(s)$ is independent of t and given by (2.29). Finally, since $K = \mathbf{T}_s\cdot\mathbf{N}$ (cf. (1.6)), $K(s)$ is also independent of t. ∎

For a steadily evolving interface that is convex, $V(\theta)$ and $K(\theta)$ are independent of time, and

$$V(\theta) = \mathbf{U}\cdot\mathbf{N}(\theta). \tag{2.30}$$

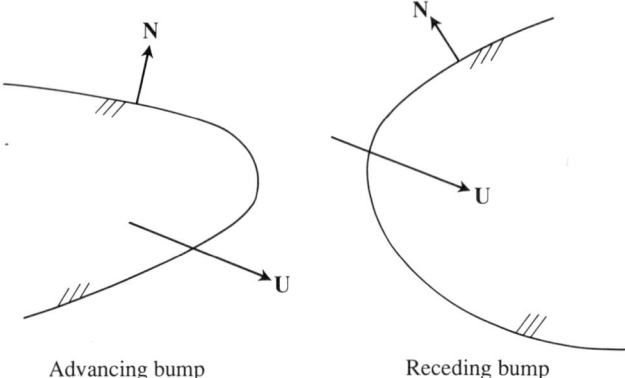

Fig. 2.1 Steadily evolving bumps. The steady velocity is given by **U**.

By a **steadily evolving bump** we mean a convex, steadily evolving interface such that

$$K(\theta)\mathbf{U}\cdot\mathbf{N}(\theta) \text{ never vanishes.} \qquad (2.31)$$

A steadily evolving bump is **advancing** or **receding** according as

$$K(\theta)\mathbf{U}\cdot\mathbf{N}(\theta) < 0 \quad \text{or} \quad K(\theta)\mathbf{U}\cdot\mathbf{N}(\theta) > 0 \qquad (2.32)$$

(Figure 2.1). Note that steadily evolving bumps are necessarily *unbounded*.

A trivial example of a steadily involving interface is a **steadily evolving facet**; these are completely determined by **U** and the corresponding normal angle θ ($\equiv$ constant).

2.5 Piecewise-smooth evolving curves

We now extend some of the previous definitions and results to evolving curves that are continuous but not smooth; as before, we write PS as an abbreviation for piecewise smooth.

A **PS evolving curve** is a finite or countably infinite list $\imath = \{\imath_1, \imath_2, \ldots\}$ of (two or more) *evolving curves* $\imath_i$, called **evolving arcs** of $\imath$, of equal duration T, such that $\imath(t) = \{\imath_1(t), \imath_2(t), \ldots\}$ is a PS curve at each $t \in [0, T)$.

We write $\mathbf{r} = \{\mathbf{r}_1, \mathbf{r}_2, \ldots\}$ for the list of arc parametrizations and refer to $\mathbf{r}$ as a **parametrization** of $\imath$. Let $[P_i(t), Q_i(t)]$ denote the parameter interval for $\imath_i$ at time t; then

$$\mathbf{r}_i(Q_i(t), t) = \mathbf{r}_{i+1}(P_{i+1}(t), t) =: \mathbf{R}_i(t) \qquad (2.33)$$

and $\mathbf{R}_i^{\cdot}(t)$ are the **position** and **total velocity of the juncture** i.

Most of the definitions concerning evolving curves extend in a natural manner to PS evolving curves.

Let $\imath$ be a PS evolving curve. An **arc-length map** for $\imath$ is a list $\{s_1, s_2, \ldots\}$ with $s_i(p, t)$ an arc-length map for $\imath_i$ and

$$s_i(Q_i(t), t) = s_{i+1}(P_{i+1}(t), t) =: S_i(t)$$

at each juncture i. It is not difficult to construct an arc-length map for $\imath$; granted one is prescribed, we can define arc length s at time t by $s = s_i(p, t)$ for any i and $p \in [Q_i(t), P_i(t)]$. This allows us to consider the tangent, normal, orientation, curvature, normal velocity, and arc velocity as functions $\mathbf{T}(s, t)$, $\mathbf{N}(s, t)$, $\theta(s, t)$, $K(s, t)$, $V(s, t)$, and $v(s, t)$ of arc length and time. These functions will generally suffer jump discontinuities across $s = S_i(t)$; with this in mind, given any function $\varphi(s, t)$, we write

$$\varphi_i^{\pm}(t) = \varphi(S_i(t) \pm 0, t). \tag{2.34}$$

We will refer to $\imath$ as **genuine** if, at each juncture i,

$$\theta_i^-(t) \neq \theta_i^+(t) \qquad \text{at some time } t. \tag{2.35}$$

We associate with each juncture i two functions of time:

$$\begin{aligned} \vartheta_i &= \theta_i^+ - \theta_i^-, \\ v_i &= \tfrac{1}{2}(V_i^+ + V_i^-); \end{aligned} \tag{2.36}$$

ϑ_i is the **juncture curvature** and v_i the **average velocity** of the juncture. Then

$$\begin{aligned} \mathbf{T}_i^+ \cdot \mathbf{N}_i^- &= -\mathbf{T}_i^- \cdot \mathbf{N}_i^+ = \sin \vartheta_i, \\ \mathbf{T}_i^+ \cdot \mathbf{T}_i^- &= \mathbf{N}_i^- \cdot \mathbf{N}_i^+ = \cos \vartheta_i. \end{aligned} \tag{2.37}$$

We will refer to $\imath$ as **convex** if its curvature K and its juncture curvatures ϑ_i never vanish, and if K and the ϑ_i *are all of the same sign*.

If $\imath$ is convex, then θ and t may be used in place of s and t as independent variables, but the juncture angles, which define the domain of θ at each t, generally vary with t. If, at each juncture i, the angles $\theta_i^-(t)$ and $\theta_i^+(t)$ are independent of t, then $\imath$ has **fixed juncture angles**.

It is convenient to write

$$v_i^{\pm} = \mathbf{T}_i^{\pm} \cdot \mathbf{R}_i^{\cdot} \tag{2.38}$$

for the **tangential endpoint velocities** at the juncture i (cf. (2.12)).

(2G) Juncture conditions *At each juncture i,*

$$\begin{aligned} V_i^- \mathbf{N}_i^- + v_i^- \mathbf{T}_i^- &= V_i^+ \mathbf{N}_i^+ + v_i^+ \mathbf{T}_i^+ = \mathbf{R}_i^{\cdot}, \\ v_i^+ - v_i^- &= \frac{2 v_i \sin \vartheta_i}{1 + \cos \vartheta_i}. \end{aligned} \tag{2.39}$$

If $\imath$ is convex with fixed juncture angles, then

$$V(\theta, t)\mathbf{N}(\theta) - V_\theta(\theta, t)\mathbf{T}(\theta) \tag{2.40}$$

is continuous across each juncture, its value at $\theta = \theta_i^\pm$ being the total velocity $\mathbf{R}_i^\cdot(t)$.

PROOF. The identities $(2.39)_1$ and (2.40) are a direct consequence of $(2.16)_1$, (2.21), and (2.33). The inner product of $(2.39)_1$ with $\mathbf{T}_i^+$ and $\mathbf{T}_i^-$ gives two equations whose difference and (2.37) yield $(2.39)_2$. ∎

Using (2.37), we can write $(2.39)_1$ in the alternative forms

$$\begin{aligned} V_i^- &= V_i^+ \cos \vartheta_i + v_i^+ \sin \vartheta_i, \\ v_i^- &= -V_i^+ \sin \vartheta_i + v_i^+ \cos \vartheta_i; \end{aligned} \tag{2.41}$$

$$\begin{aligned} V_i^+ &= V_i^- \cos \vartheta_i - v_i^- \sin \vartheta_i, \\ v_i^+ &= V_i^- \sin \vartheta_i + v_i^- \cos \vartheta_i. \end{aligned} \tag{2.42}$$

A **fully faceted evolving curve** is a PS evolving curve $\imath = \{\imath_1, \imath_2, \ldots\}$ each of whose arcs is a *facet*. If, in addition, $\theta(s, t)$ is independent of s and t on each facet, then $\imath$ has **fixed normal angles**. An example of a fully faceted evolving curve with fixed normal angles is a **wrinkling**; here there are constant angles α and β, the *normal angles of the wrinkling*, such that, for all t,

$$\begin{aligned} \theta(s, t) &= \alpha &&\text{on } \imath_i(t) \text{ for } i \text{ odd,} \\ \theta(s, t) &= \beta &&\text{on } \imath_i(t) \text{ for } i \text{ even.} \end{aligned} \tag{2.43}$$

Let $\imath$ be a fully faceted evolving curve with fixed normal angles. Then, by $(2.9)_2$, $V_s = 0$ on each facet, so that $V = V(t)$ on each facet. In this case we will denote the normal angle and normal velocity of $\imath_i$ by θ_i and $V_i(t)$. By the agreement (2.34), $\theta_i^\pm$ and $V_i^\pm(t)$ designate the limits at the juncture i of the normal angle and normal velocity, so that

$$\begin{aligned} \theta_i^- &= \theta_{i-1}, &\theta_i^+ &= \theta_i, \\ V_i^-(t) &= V_{i-1}(t), &V_i^+(t) &= V_i(t), \end{aligned} \tag{2.44}$$

and the juncture curvature of i is given by

$$\vartheta_i = \theta_i - \theta_{i-1}. \tag{2.45}$$

We denote the **length** of $\imath_i$ by $L_i(t)$.

(2H) Kinematics of fully faceted evolving curves with fixed normal angles *A fully faceted evolving curve with fixed normal angles evolves according to*

$$L_i^{\cdot} = (\cot \vartheta_{i+1} + \cot \vartheta_i)V_i - (\sin \vartheta_i)^{-1}V_{i-1} - (\sin \vartheta_{i+1})^{-1}V_{i+1}. \quad (2.46)$$

at each juncture i.

PROOF. If we apply $(2.41)_1$ to the juncture i and $(2.42)_1$ to the juncture $i+1$, and use the agreements (2.44), we find that

$$V_{i-1} = V_i \cos \vartheta_i + (v_i^+) \sin \vartheta_i,$$
$$V_{i+1} = V_i \cos \vartheta_{i+1} - (v_{i+1}^-) \sin \vartheta_{i+1}.$$

Further, by (2.18),

$$L_i^{\cdot} = (v_{i+1}^-) - (v_i^+).$$

These equations, when combined, yield (2.46). ∎

(2I) Transport theorem for integrals *Let φ be a piecewise-continuous, piecewise-smooth function on a PS evolving curve $\imath$, with φ smooth on each arc of $\imath$. Then*

$$\frac{d}{dt}\int_{\imath(t)} \varphi \, ds = \int_{\imath(t)} (\varphi^{\circ} - \varphi KV) \, ds + \int_{\partial \imath(t)} \varphi v_{\partial \imath(\tan)} - \sum (\varphi_i^+ v_i^+ - \varphi_i^- v_i^-),$$
$$(2.47)$$

where the sum is over the essential junctures i of $\imath$.

PROOF. We apply the transport theorem (2D) to each of the arcs of $\imath$ and then add the resulting relations. ∎

Let φ and ψ be piecewise-continuous, piecewise-smooth functions on a closed PS evolving curve $\imath$, with φ and ψ smooth on each arc of $\imath$. Then integrating $\varphi \psi_s$ by parts on each arc and then adding the resulting relations yields the following formula for integration by parts:

$$\int_{\imath(t)} \varphi \psi_s \, ds = -\int_{\imath(t)} \varphi_s \psi \, ds - \sum (\varphi_i^+ \psi_i^+ - \varphi_i^- \psi_i^-). \quad (2.48)$$

(2J) Transport theorem for area and perimeter *Let $\imath$ be a closed, simple PS evolving curve. Then using the notation of (2E) and (2I),*

$$A^{\cdot}(t) = \int_{\imath(t)} V \, ds.$$
$$L^{\cdot}(t) = -\int_{\imath(t)} KV \, ds + \sum (v_i^+ - v_i^-),$$
$$(2.49)$$

PROOF. The identity $(2.49)_2$ follows from (2.47). The verification of $(2.49)_1$ is analogous to that of $(2.26)_1$. By (1.16) (which also holds for PS curves), (2.7), (2.27), and (2.47),

$$2A^{\cdot}(t) = \int_{\imath(t)} (V - \mathbf{r}\cdot\mathbf{T}V_s - KV\mathbf{r}\cdot\mathbf{N})\,ds - \sum [(\mathbf{r}\cdot\mathbf{N})_i^+ v_i^+ - (\mathbf{r}\cdot\mathbf{N})_i^- v_i^-].$$

On the other hand, by (2.48), (1.2), and (1.6),

$$-\int_{\imath(t)} \mathbf{r}\cdot\mathbf{T}V_s\,ds = \int_{\imath(t)} (V + KV\mathbf{r}\cdot\mathbf{N})\,ds + \sum [(\mathbf{r}\cdot\mathbf{T})_i^+ V_i^+ - (\mathbf{r}\cdot\mathbf{T})_i^- V_i^-].$$

Let $\mathbf{Q}$ be the orthogonal transformation that transforms $\mathbf{T}$ into $\mathbf{N}$ and $\mathbf{N}$ into $-\mathbf{T}$. Then $\mathbf{r}\cdot\mathbf{N} = -\mathbf{Q}\mathbf{r}\cdot\mathbf{T}$ and $\mathbf{r}\cdot\mathbf{T} = \mathbf{Q}\mathbf{r}\cdot\mathbf{N}$, so that, by (2.39), the last two identities combine to give

$$2A^{\cdot}(t) = \int_{\imath(t)} 2V\,ds + \sum \mathbf{R}_i^{\cdot} \cdot [(\mathbf{Q}\mathbf{r})_i^+ - (\mathbf{Q}\mathbf{r})_i^-],$$

which implies the desired result, since $\mathbf{Qr}$ is continuous across each juncture. ∎

2.6 Variational lemmas

Certain proofs require evolving curves with specific properties. The next two lemmas establish the existence of such curves.

(2K) First variational lemma *Let $\imath_0$ be a (fixed) curve.*

(a) *Let $[S_1, S_2]$ denote the arc-length interval for $\imath_0$. Let ω be a smooth function on $[S_1, S_2]$ with compact support in (S_1, S_2). Then there is an evolving curve $\imath(t), 0 \leq t < T$, with $\imath(0) = \imath_0$, whose normal velocity $V(s, t)$ satisfies*

$$V(s, 0) = \omega(s) \qquad (2.50)$$

for $S_1 \leq s \leq S_2$.

(b) *Let $\mathbf{r}_0$ be an endpoint of $\imath_0$. Then given a vector $\mathbf{a}$ there is an evolving curve $\imath(t), 0 \leq t < T$, with $\imath(0) = \imath_0$, such that the endpoint $\mathbf{R}(t)$ with $\mathbf{R}(0) = \mathbf{r}_0$ satisfies $\mathbf{R}^{\cdot}(t) = \mathbf{a}$ for all t, while a neighbourhood of the other endpoint is motionless.*

PROOF.

(a) Let $\mathbf{r}_0(s)$ be an arc-length parametrization of $\imath_0$, let $\mathbf{N}_0(s)$ denote the normal to $\imath_0$, and define

$$\mathbf{r}(p, t) = \mathbf{r}_0(p) + t\omega(p)\mathbf{N}_0(p).$$

Since $|\mathbf{r}_p(p, 0)|$ is bounded away from zero on $[S_1, S_2]$, for some $T > 0$, $\mathbf{r}(p, t)$, $S_1 \leq p \leq S_2$, $0 \leq t < T$, parametrizes an evolving curve $\imath(t)$, and the normal velocity of $\imath(t)$ has the desired properties.

(b) Assume that $\mathbf{r}_0$ is the terminal point of $\imath_0$. Let $\omega(s)$ be a smooth increasing function on $[S_1, S_2]$ that vanishes on $[S_1, S_0]$, $S_0 \in (S_1, S_2)$, and has $\omega(S_2) = 1$. Then for some $T > 0$,

$$\mathbf{r}(p, t) = \mathbf{r}_0(p) + t\omega(p)\mathbf{a},$$

$S_1 \leq p \leq S_2$, $0 \leq t < T$, parametrizes an evolving curve $\imath(t)$ with the desired properties. The proof for $\mathbf{r}_0$ the initial point is analogous. ∎

(2L) Second variational lemma *Let θ_0, θ_1, V_0, V_1, K_0, K_1, $T_0 > 0$, and T_1 be prescribed scalars. Then there is an evolving interface $\imath$ (with parametrization $\mathbf{r}$, normal angle θ, normal velocity V, and curvature K), and a smooth, strictly positive field T on $\mathbb{R}^2 \times [0, \infty)$ such that, at some (p, t) in the domain of $\mathbf{r}$:*

$$\theta = \theta_0, \qquad \theta^\circ = \theta_1, \qquad V = V_0, \qquad V^\circ = V_1,$$
$$K = K_0, \qquad K^\circ = K_1, \qquad T_\imath = T_0, \qquad T_\imath^\circ = T_1,$$

with $T_\imath(q, t) = T(\mathbf{r}(q, t), t)$.

Exercise Prove (2L). Hint: let φ be a smooth function on $[0, \infty)$ with $\varphi(0) = 0$, $\varphi\dot{}(0) = T_1$, and $T_0 + \varphi(t) > 0$ for all $t \geq 0$; show that $\mathbf{r}$ and T defined by

$$\mathbf{r}(p, t) = (p, h(p, t)),$$
$$h(p, t) = V_0 t + \tfrac{1}{2}V_1 t^2 + \tfrac{1}{2}K_0 p^2 + \tfrac{1}{2}tK_1 p^2,$$
$$T(\mathbf{x}, t) = T_0 + \varphi(t),$$

have the requisite properties.

3
PHASE REGIONS, CONTROL VOLUMES, AND INFLOWS

3.1 Phase regions and control volumes

We consider an **evolving interface** $\mathscr{s}$ separating two phases, labelled 1 and 2, and write $\Omega_1(t)$ and $\Omega_2(t)$ for the corresponding **phase regions**, the sub-regions[10] of the body occupied by phase 1 and phase 2. We assume that the body occupies all of $\mathbb{R}^2$, and that $\Omega_1(t)$ and $\Omega_2(t)$ are closed regions with $\mathbb{R}^2$ as union and $\mathscr{s}(t)$ as intersection. Further, (without loss in generality) we choose the normal $\mathbf{N}$ of $\mathscr{s}$ so that

$$\Omega_1(t) \text{ is the reference region for } \mathscr{s}(t); \tag{3.1}$$

$\mathbf{N}(s, t)$ then coincides with the outward unit normal to $\partial\Omega_1(t)$.

Let $R \subset \mathbb{R}^2$ be a bounded region with piecewise-smooth boundary. Then

$$\mathscr{i}(t) = R \cap \mathscr{s}(t), \tag{3.2}$$

is the portion of the interface that lies in R, while

$$R_1(t) = R \cap \Omega_1(t), \qquad R_2(t) = R \cap \Omega_2(t) \tag{3.3}$$

are the portions of the individual phase regions that lie in R. We will refer to R as a **control volume** during a time interval $\mathscr{T} = \mathscr{T}(R)$ if, for all $t \in \mathscr{T}$, either:

(1) is an evolving subcurve of $\mathscr{s}$, and $R_1(t)$ and $R_2(t)$ are non-empty regions; or

(2) R lies in the interior of one of the phase regions during $\mathscr{T}$.

In case (1) R is a **two-phase control volume** (Figure 3.1); in (2) R is a **bulk control volume**. We will generally omit the argument t when writing integrals involving $R_1(t)$, $R_2(t)$, $\mathscr{i}(t)$, or $\partial\mathscr{i}(t)$, and in all assertions concerning a control volume R, it will be tacit that t is restricted to the underlying time interval $\mathscr{T}$.

3.2 Inflows, the pillbox lemma, and infinitesimally thin evolving control volumes

We will discuss two types of fields: **bulk fields** defined in each of the bulk regions for all time; and **interfacial fields** defined on the interface for all time.

---

[10] We use the term **region** as a synonym for connected open set in $\mathbb{R}^2$.

PHASE REGIONS, CONTROL VOLUMES, AND INFLOWS

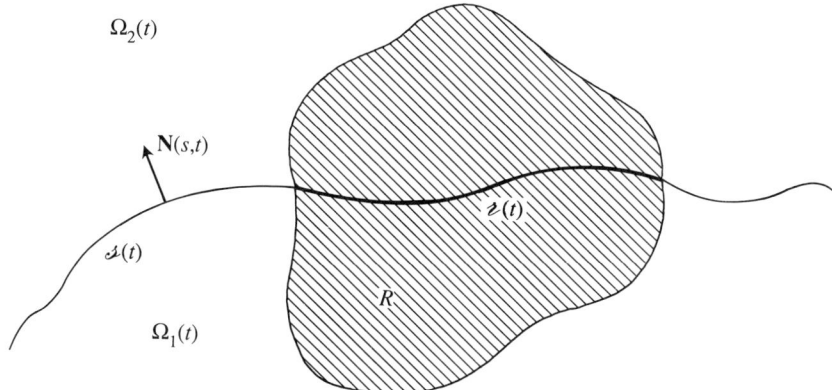

Fig. 3.1 A two-phase control volume R.

We will not specify regularity hypotheses other than to note that *bulk fields are allowed to suffer jump discontinuities across the interface*. For φ a bulk field, we write $[\varphi]$ for the **jump** in φ **across the interface**: for $\mathbf{x} \in \mathcal{J}(t)$,

$$[\varphi](\mathbf{x}, t) = \lim_{\substack{\mathbf{y} \to \mathbf{x} \\ \mathbf{y} \in \Omega_2(t)}} \varphi(\mathbf{y}, t) - \lim_{\substack{\mathbf{y} \to \mathbf{x} \\ \mathbf{y} \in \Omega_1(t)}} \varphi(\mathbf{y}, t). \tag{3.4}$$

A standard identity then yields (cf. $(2.26)_1$)

$$\frac{d}{dt} \int_R \varphi \, da = \int_{R_1} \varphi^\bullet \, da + \int_{R_2} \varphi^\bullet \, da - \int_{\imath} [\varphi] V \, ds. \tag{3.5}$$

Let $\imath$ be an evolving subcurve of $\mathcal{J}$. It is possible to construct a two-phase control volume R such that $\imath(t)$ is the portion of the interface in R at $t = t_0$; in fact, it is possible to construct a sequence of such control volumes that shrinks to $\imath(t_0)$. Fix $t_0 > 0$, let $[S_1, S_2]$ denote the interval of arc lengths for $\imath(t_0)$, and, for $\varepsilon > 0$, let

$$R(\varepsilon) = \{\mathbf{r}(s, t_0) + \lambda \mathbf{N}(s, t_0) : 0 \leq |\lambda| < \varepsilon, S_1 < s < S_2\}.$$

Then there is a time interval $\mathcal{T}$ containing t_0 such that, for all sufficiently small ε, say $0 < \varepsilon < \varepsilon_0$:

(1) $R(\varepsilon)$ is a two-phase control volume during $\mathcal{T}$;

(2) $\imath(t_0)$ is the portion of the interface in $R(\varepsilon)$ at t_0.

The family $R(\varepsilon)$, $0 < \varepsilon < \varepsilon_0$, will be referred to as a **pillbox that shrinks to** $\imath(t_0)$ (Figure 3.2).

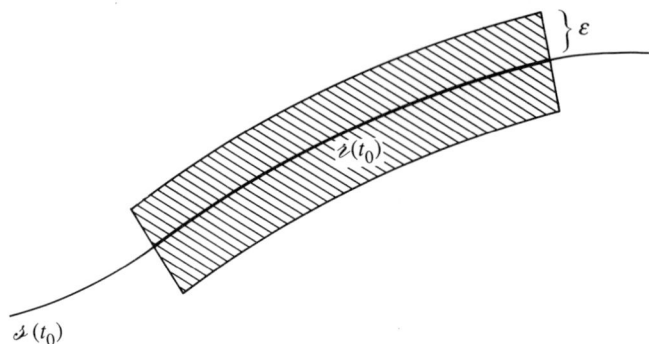

Fig. 3.2 A pillbox that shrinks to $\imath(t_0)$.

(3A) Pillbox lemma *Let φ and $\mathbf{f}$ be bulk fields with φ scalar-valued and $\mathbf{f}$ vector-valued. Let $\imath$ be an evolving subcurve of $\mathfrak{s}$, and let $R(\varepsilon)$, $0 < \varepsilon < \varepsilon_0$, be a pillbox that shrinks to $\imath(t_0)$. Then, as $\varepsilon \to 0$,*

$$\left\{\frac{d}{dt}\int_{R(\varepsilon)} \varphi \, da\right\}_{t=t_0} \to -\int_{\imath(t_0)} [\varphi]V \, ds,$$
$$\left\{\int_{\partial R(\varepsilon)} \mathbf{f}\cdot\mathbf{n} \, ds\right\}_{t=t_0} \to \int_{\imath(t_0)} [\mathbf{f}]\cdot\mathbf{N} \, ds \qquad (3.6)$$

PROOF. The limit $(3.6)_1$ follows from (3.5), since the area of $R(\varepsilon)$ tends to zero as ε tends to zero. To establish $(3.6)_2$, let $\Gamma_1(\varepsilon)$ and $\Gamma_2(\varepsilon)$ denote the portions of $\partial R(\varepsilon)$ in Ω_1 and Ω_2 at $t = t_0$. Then, at $t = t_0$,

$$\int_{\Gamma_1(\varepsilon)} \mathbf{f}\cdot\mathbf{n} \, da \to -\int_{\imath(t_0)} \mathbf{f}\cdot\mathbf{N} \, da, \qquad \int_{\Gamma_2(\varepsilon)} \mathbf{f}\cdot\mathbf{n} \, da \to \int_{\imath(t_0)} \mathbf{f}\cdot\mathbf{N} \, da$$

as $\varepsilon \to 0$, and $(3.6)_2$ follows. ∎

Let $\imath$ be an evolving subcurve of $\mathfrak{s}$. When discussing balance laws it is useful to visualize $\imath(t)$ as an infinitesimally thin region consisting of:

(1) the portion of the interface in $\imath(t)$;

(2) the points of the bulk material immediately adjacent to $\imath(t)$.

With this interpretation we will refer to $\imath$ as an **infinitesimally thin two-phase evolving control volume** (Figure 3.3). The **physical boundary** of $\imath(t)$ then consists of:

(1) the edges $\partial\imath(t)$, which represent the interaction of $\imath(t)$ with the remainder of the interface;

(2) the *two sides* of $\imath(t)$, which represent the interaction of $\imath(t)$ with the bulk material.

PHASE REGIONS, CONTROL VOLUMES, AND INFLOWS

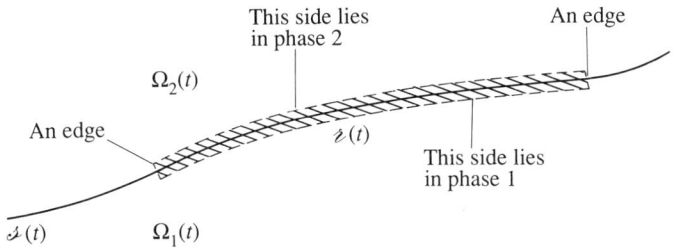

Fig. 3.3 An infinitesimally thin two-phase evolving control volume.

These definitions are formal and we will use them only as aids when writing balance laws; *in mathematical expressions, such as integrals, $\imath$ should always be interpreted with its precise mathematical meaning as an evolving subcurve.*

Let φ be a bulk field that represents the bulk density of a physical quantity, which, for convenience, we will refer to as the energy, and let $\varphi_1(\mathbf{x}, t)$ and $\varphi_2(\mathbf{x}, t)$ denote the individual limits in (3.4) from $\Omega_1(t)$ and $\Omega_2(t)$. Consider an infinitesimally thin two-phase evolving control volume $\imath$. The bulk energy of $\imath$ is zero, since $\imath$ is infinitesimally thin, but, since $\imath$ is moving relative to the bulk material, $\imath$ captures and loses bulk energy across its physical boundary. In fact,

$$-\int_{\imath} \varphi_1 V \, ds, \quad \int_{\imath} \varphi_2 V \, ds$$

represent *inflows* of bulk energy from phases 1 and 2; thus

$$\int_{\imath} [\varphi] V \, ds \tag{3.7}$$

represents the total inflow of bulk energy to the control volume $\imath$. Similarly, if $\mathbf{f}$ is a bulk vector field that represents, for example, the heat flux, then

$$\int_{\imath} \mathbf{f}_1 \cdot \mathbf{N} \, ds, \quad -\int_{\imath} \mathbf{f}_2 \cdot \mathbf{N} \, ds$$

represent heat flows into the control volume from phases 1 and 2, and thus

$$-\int_{\imath} [\mathbf{f}] \cdot \mathbf{N} \, ds \tag{3.8}$$

gives the net heat flow into the control volume $\imath$ from the bulk material.

Let $R(\varepsilon)$ be a pillbox that shrinks to $\imath(t_0)$. Then the result (3.6) asserts that the rate at which the bulk energy of $R(\varepsilon)$ is changing tends to the outflow of bulk energy from any infinitesimally thin evolving control volume that coincides with $\imath(t_0)$ at time t_0, and the heat flow into the pillbox from the bulk material tends to the heat flow into any such evolving control volume.

The following lemma will be useful in establishing local balance laws.

(3B) Localization lemma *Let φ and ω be interfacial fields and suppose that*

$$\int_\imath \varphi \, ds + \int_{\partial \imath} \omega v_{\partial\imath(\tan)} \leq 0 \tag{3.9}$$

for all evolving subcurves $\imath$ of $\mathfrak{s}$. Then $\varphi \leq 0$, $\omega = 0$. If '$\leq$' is replaced by '$=$' in (3.9), then $\varphi = 0$.

PROOF. Choose a time $\tau \in [0, T)$ ($T =$ duration of $\mathfrak{s}$), an arc-length interval $[s_1, s_2]$ for $\imath(\tau)$, and scalars α_1 and α_2. Then there are smooth functions $S_1(t)$ and $S_2(t)$ on $[0, T)$ such that $S_1(t) < S_2(t)$ and $S_i(\tau) = s_i$, $S_i'(\tau) = \alpha_i + v(s_i \tau)$ ($i = 1, 2$), where v is the arc velocity for $\mathfrak{s}$ (cf. (2.2)). Then $[S_1(t), S_2(t)]$ is an arc-length interval for an evolving subcurve $\imath(t)$ of $\mathfrak{s}$, so that, by (2.15), (3.9) yields

$$\int_{\imath(\tau)} \varphi \, ds + \omega(s_2, \tau)\alpha_2 - \omega(s_1, \tau)\alpha_1 \leq 0.$$

Since τ, s_1, s_2, α_1, and α_2 were arbitrarily chosen, $\omega \equiv 0$; the remainder of the proof is elementary. ∎

II
MECHANICAL THEORY OF INTERFACIAL EVOLUTION

4
BALANCE OF FORCES; POWER

4.1 Balance of forces[11]

Consider an evolving interface $\mathcal{A}$ with $\mathbf{r}$ a corresponding parametrization. The micromechanics of the interface is described by three interfacial fields:

$\mathbf{C}(s, t)$ capillary force,

$\mathbf{B}(s, t)$ interactive force,

$m(s, t)$ interactive moment.

$\mathbf{C}(s, t)$ represents the force *within* the interface and generalizes the classical notion of surface tension. In fact, if we decompose this vector into tangential and normal components,

$$\mathbf{C} = \sigma \mathbf{T} + \xi \mathbf{N}, \tag{4.1}$$

then $\sigma(s, t)$ represents the **surface tension**, while $\xi(s, t)$ is the **surface shear**. In classical continuum mechanics infinitesimally thin rods have no associated internal shearing forces. Here, as we shall see, even though the interface has no thickness, shearing forces are required whenever anisotropy is present.

The vector $\mathbf{B}(s, t)$ and the scalar $m(s, t)$ represent the force and moment, per unit interfacial length, exerted at the interface by the bulk material.

Consider an evolving subcurve $\imath$ interpreted as an infinitesimally thin two-phase evolving control volume. Then **balance of forces and moments** for $\imath$ is the requirement (Figure 4.1) that[12]

$$\int_{\partial \imath} \mathbf{C} + \int_{\imath} \mathbf{B} \, ds = 0,$$
$$\int_{\partial \imath} \mathbf{r} \times \mathbf{C} + \int_{\imath} \mathbf{r} \times \mathbf{B} \, ds + \int_{\imath} m \, ds = 0. \tag{4.2}$$

These laws must hold for all such subcurves $\imath$; we therefore conclude, with the aid of (1.15), that

$$\mathbf{C}_s + \mathbf{B} = 0,$$
$$\mathbf{r}_s \times \mathbf{C} + m = 0, \tag{4.3}$$

---

[11] Cf. Gurtin (1988).
[12] For $\mathbf{a} = a_1 \mathbf{T} + a_2 \mathbf{N}$ and $b = b_1 \mathbf{T} + b_2 \mathbf{N}$, $\mathbf{a} \times \mathbf{b} = a_1 b_2 - b_1 a_2$ (a scalar).

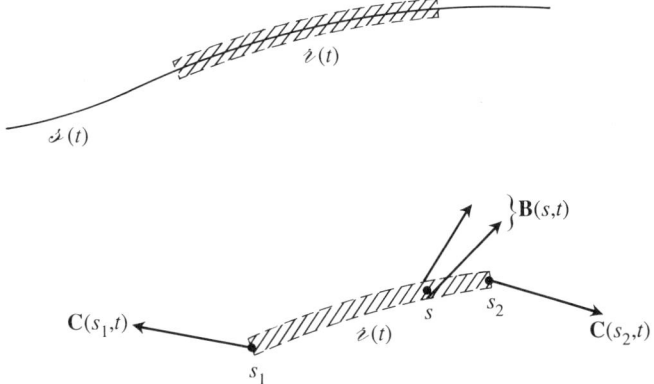

Fig. 4.1 Forces on an evolving subcurve $\imath(t)$.

or equivalently, by (1.2), (1.6), (1.8), and (4.1),

$$\xi_s + \sigma K + B_{\text{norm}} = 0, \qquad \sigma_s - \xi K + B_{\text{tan}} = 0, \qquad \xi = -m. \qquad (4.4)$$

(4A) Remark A direct consequence of (4.2)$_1$ is that, for each t,

$$\mathbf{C}(s, t) \text{ is continuous in } s. \qquad (4.5)$$

This result will yield an important free-boundary condition when we discuss interfacial curves with corners.

4.2 The power identity

Motion of the interface should involve an expenditure of power. In mechanics the power expended on a particle is generally the force on the particle times the particle velocity. There is, however, no canonical description of the interface as a system of particles, chiefly because the material points that comprise the interface change from time to time. In fact, interfacial power should depend on kinematical quantities intrinsic to the portion of the interface under consideration.

Let $\imath$ be an evolving subcurve of $\jmath$. Power is expended on $\imath(t)$ by the portion of the interface exterior to $\imath(t)$ through the action of the capillary force $\mathbf{C}$ on the endpoints of $\imath(t)$. Power is also expended on $\imath(t)$ by the bulk material through the interactive force $\mathbf{B}$ and moment m acting along the length of $\imath(t)$. We assume that $\mathbf{C}$ expends power over the corresponding endpoint velocity $\mathbf{v}_{\partial \imath}$ with total power expense given by

$$\int_{\partial \imath} \mathbf{C} \cdot \mathbf{v}_{\partial \imath}. \qquad (4.6)$$

Similarly, we assume that **B** expends power over the normal velocity $V\mathbf{N}$, while m expends power over the angular velocity $\theta°$; thus

$$\int_{\imath} (\mathbf{B} \cdot V\mathbf{N} + m\theta°)\, ds \qquad (4.7)$$

represents the power expended on $\imath(t)$ by the bulk material.

By (2.23) and (4.1),

$$\int_{\partial \imath} \mathbf{C} \cdot \mathbf{v}_{\partial \imath} = \int_{\partial \imath} \sigma v_{\partial \imath (\text{tan})} + \int_{\partial \imath} \xi V, \qquad (4.8)$$

and hence capillary power manifests itself through the action of surface tension as material is added to the edge of $\imath(t)$ and through the action of surface shear as $\imath(t)$ moves normally.

The next result is central to the thermodynamics of phase transitions.

(4B) Power identity[13] *Given any evolving subcurve $\imath$ of $\jmath$.*

$$\int_{\partial \imath} \mathbf{C} \cdot \mathbf{v}_{\partial \imath} + \int_{\imath} \mathbf{B} \cdot V\mathbf{N}\, ds + \int_{\imath} m\theta°\, ds = -\int_{\imath} \sigma K V\, ds + \int_{\partial \imath} \sigma v_{\partial \imath (\text{tan})}. \qquad (4.9)$$

The identity (4.9) asserts that *all of the power expended on an evolving subcurve goes into the creation of interface*: at internal points through $-\sigma K V$ and at the endpoints through $\sigma v_{\partial \imath (\text{tan})}$.

PROOF. By $(2.9)_2$ and (4.4),

$$\int_{\partial \imath} \xi V = \int_{\imath} (\xi_s V + \xi V_s)\, ds = -\int_{\imath} (\mathbf{B} \cdot V\mathbf{N} + \sigma K V + m\theta°)\, ds;$$

this relation and (4.8) imply (4.9). ∎

(4C) Remark We have taken the *normal velocity* as the kinematical variable over which the interactive force expends power; the tangential force B_{tan} does not perform work. As is customary with a 'constraint' of this type, we leave B_{tan} as *indeterminate*: when we discuss constitutive equations, we will postulate a constitutive equation for B_{norm}, but none for B_{tan}.

[13] Cf. Gurtin (1988).

5
ENERGETICS AND THE DISSIPATION INEQUALITY[14]

We consider a purely mechanical theory in which the interface is driven by an energy difference between bulk phases. We assume that the body is immobile, incapable of deformation, with energy uniform in each of the bulk phases. We base the theory on a *dynamical* version of the second law, called the *dissipation inequality*, which requires that:

> The energy of a control volume R changes at a rate not greater than the power expended on R.

The energetics is described by two fields:

(1) a bulk field

$$\psi(x, t) = \begin{cases} \psi_1, & x \in \Omega_1(t) \\ \psi_2, & x \in \Omega_2(t), \end{cases} \tag{5.1}$$

where ψ_1 and ψ_2 are the *constant* **bulk energies** of phase 1 and phase 2;

(2) an interfacial field $f(s, t)$, which represents the **energy of the interface**.

The total energy of a two-phase control volume is then given by

$$\int_R \psi \, da + \int_\iota f \, ds, \tag{5.2}$$

where $\iota(t)$ is the portion of the interface that lies in R (cf. (3.2)).

The dissipation inequality involves power expended on R. Guided by the power identity (4.9), we assume that the portion of the *interface* exterior to R expends power on R through the action of capillary forces on $\partial \iota$, with this expenditure given by (4.6). This represents the sole source of power for R: bulk forces acting on ∂R do not expend power since the bulk material is immobile; the interactive force **B** and moment m do not expend power on R since they are internal to R.

Within this purely mechanical framework the **second law** has the form of the **dissipation inequality**

$$\frac{d}{dt} \left\{ \int_R \psi \, da + \int_\iota f \, ds \right\} \leq \int_{\partial \iota} \mathbf{C} \cdot \mathbf{v}_{\partial \iota}, \tag{5.3}$$

which is required to hold for all two-phase control volumes R.

---
[14] Cf. Gurtin (1988, 1991); Angenent and Gurtin (1989); Gurtin and Struthers (1990).

Choose an evolving subcurve $\imath$ of $\mathfrak{s}$ and an arbitrary time t_0, apply (5.3) to a pillbox $R(\varepsilon)$, $0 < \varepsilon < \varepsilon_0$, that shrinks to $\imath(t_0)$, let $\varepsilon \to 0$, and use (3.6) and the fact that t_0 is arbitrary; the result is the **interfacial dissipation inequality**

$$\frac{d}{dt}\int_{\imath} f \, ds \leq \int_{\imath} [\![\psi]\!] V \, ds + \int_{\partial \imath} \mathbf{C} \cdot \mathbf{v}_{\partial \imath}, \tag{5.4}$$

which is valid *for all evolving subcurves $\imath$ of $\mathfrak{s}$*.

The interfacial dissipation inequality (5.4) represents the second law for a subcurve $\imath$ visualized as an infinitesimally thin two-phase evolving control volume: the first term is the rate at which the interfacial energy of $\imath(t)$ is changing;

$$\int_{\imath} [\![\psi]\!] V \, ds \tag{5.5}$$

gives the inflow of bulk energy arising from the motion of $\imath(t)$; the last term in (5.4) is the power expended on $\imath(t)$ by the material exterior to $\imath(t)$. We could equally well base the theory on (5.4). In fact, we will use such a formulation in Section 13.2.

If we apply the transport theorem (2.24) to (5.4), and use (4.4)$_3$ and the power identity (4.9), we find that

$$\int_{\imath} \{f^\circ - \xi\theta^\circ - (f - \sigma)KV + (B_{\text{norm}} - [\![\psi]\!])V\} \, ds + \int_{\partial \imath} \{f - \sigma\} v_{\partial \imath(\text{tan})} \leq 0.$$
$$\tag{5.6}$$

This inequality must hold for all evolving subcurves $\imath$ of $\mathfrak{s}$; thus, by Lemma 3B, we have the tension–energy theorem.

(5A) Tension–energy theorem *The surface tension and the interfacial energy coincide*:

$$\sigma = f. \tag{5.7}$$

If we substitute (5.7) into (5.6) and use the fact that $\imath$ is arbitrary, we are led to the **reduced dissipation inequality**

$$f^\circ - \xi\theta^\circ + (F + B)V \leq 0, \tag{5.8}$$

where we have written

$$B = B_{\text{norm}} \tag{5.9}$$

for the **normal interaction** and

$$F = -[\![\psi]\!] \tag{5.10}$$

for the constant difference in bulk energies. The inequality (5.8) plays a central role in determining appropriate constitutive equations for the interface.

Exercise Consider a *piecewise-smooth* evolving interface $\mathscr{a}$. Assume that f is smooth on each evolving arc of $\mathscr{a}$. Show that, granted (4.5), if (5.4) holds for every evolving arc of $\mathscr{a}$, then (5.4) holds for every evolving subcurve of $\mathscr{a}$.

(5B) Remark In Chapter 4 we considered **B** to be the force $\mathbf{B} = \mathbf{B}_{int}$ exerted on the interface by the bulk material. We could generalize the theory[15] by allowing **B** to include 'body forces' $\mathbf{B}_{ext}$ exerted on the interface by the external world: $\mathbf{B} = \mathbf{B}_{int} + \mathbf{B}_{ext}$. In this case we would add the term

$$\int_{\mathscr{a}} \mathbf{B}_{ext} \cdot V\mathbf{N} \, ds$$

to the right side of (5.4). The results (5.7) and (5.8) would then remain valid with $B = B_{norm}$ the normal component of $\mathbf{B}_{int}$.

Exercise Establish the assertion made in the last sentence of (5B).

---

[15] Cf. Angenent and Gurtin (1989), Section 3.1.

6
CONSTITUTIVE THEORY[16]

6.1 Constitutive equations and the compatibility theorem

As **constitutive equations** we allow the energy, the capillary force, and the normal interaction to depend smoothly on the orientation of the interface through a dependence on θ and on the kinetics of the interface through a dependence on V:

$$f = f(\theta, V), \qquad \mathbf{C} = \mathbf{C}(\theta, V), \qquad \mathbf{B} = \mathbf{B}(\theta, V). \tag{6.1}[17]$$

The first two relations characterize the interface, and the last models the interaction between the interface and the bulk material. By (1.3), (4.1), and (4.4)$_3$, the constitutive equation for the capillary force implies corresponding equations for the surface tension, surface shear, and interactive moment:

$$\sigma = \sigma(\theta, V), \qquad \xi = \xi(\theta, V), \qquad m = m(\theta, V); \tag{6.2}$$

in fact, (4.4)$_3$ and (5.7) imply that

$$\sigma(\theta, V) = f(\theta, V), \qquad m(\theta, V) = -\xi(\theta, V). \tag{6.3}$$

We will assume that the jump F in bulk energies is prescribed.

Given an evolving interface $\mathfrak{s}$, the constitutive equations may be used to compute a corresponding **constitutive process** $(f, \mathbf{C}, \mathbf{B})$; without suitable restrictions on the constitutive equations such processes will generally not all satisfy the reduced dissipation inequality (5.8). When all constitutive processes are consistent with (5.8) we will refer to the constitutive equations as **compatible with thermodynamics**.[18]

(6A) Compatibility theorem[19] *The constitutive equations are compatible with thermodynamics if and only if:*

---

[16] Cf. Gurtin (1988); Angenent and Gurtin (1989).
[17] With each of the constitutive functions 2π-periodic in θ. The equation $f = f(\theta, V)$ might, less ambiguously, be written in the form $f = \hat{f}(\theta, V)$ signifying that $f(s, t) = \hat{f}(\theta(s, t), V(s, t))$ for all s and t.
[18] The requirement that constitutive processes be consistent with (5.8) may be justified as follows. Consider the theory generalized as in Remark 5B to include external body forces $\mathbf{B}_{\text{ext}}$. Choose a constitutive process. Then, granted (4.4)$_3$, balance of moments is satisfied, while balance of forces is satisfied provided we choose the external body force appropriately. Thus all of the basic laws of the theory, except for the dissipation inequality, are satisfied. Compatibility with thermodynamics ensures that all processes be consistent with the dissipation inequality.
[19] This is a special case of a more general theorem in which the bulk material is allowed to conduct heat (cf. Theorem 16B).

(a) *the energy, surface tension, and surface shear are independent of V and satisfy*

$$\sigma(\theta) = f(\theta), \qquad \xi(\theta) = f'(\theta); \tag{6.4}$$

(b) *the normal interaction has the form*

$$B(\theta, V) = -F - b(\theta, V)V,$$
$$b(\theta, V) \geq 0. \tag{6.5}$$

PROOF. The following result will be useful:

$$\begin{aligned}&\text{Let } \phi(z) \text{ be smooth and satisfy } \phi(z)z \geq 0\\&\text{for all } z \in \mathbb{R}. \text{ Then there is a smooth function}\\&\mu(z) \geq 0 \text{ such that } \phi(z) = \mu(z)z.\end{aligned} \tag{6.6}$$

The proof is simple: $\phi(z)z$ has a minimum at $z = 0$; thus $\phi(0) = 0$, so that $\mu(z) = z^{-1}\phi(z) \geq 0$ is well defined and smooth at $z = 0$.

To prove the theorem, we note that, in view of the constitutive equations, (5.8) is equivalent to the inequality

$$[f_\theta(\theta, V) - \xi(\theta, V)]\theta^\circ + f_V(\theta, V)V^\circ + [B(\theta, V) + F]V \leq 0. \tag{6.7}$$

Assume that (6.7) holds for all evolving interfaces. Then, by Lemma (2L), we can construct an interfacial motion for which, at some point and time, the fields θ, V, θ°, and V° have arbitrary values; this implies (a) and the inequality

$$[F + B(\theta, V)]V \leq 0, \tag{6.8}$$

and, appealing to (6.6) with $z = V$ and $\phi(z) = -F - B(\theta, z)$, assertion (b) follows.

Conversely, (a) and (b) trivially yield (6.7) in all processes. ∎

We will refer to $b(\theta, V)$ as the **kinetic coefficient**.

(6B) Remarks

(1) The normal interaction B may be regarded as a force in the direction of V exerted on the interface by the bulk material. Equation (6.5) gives this force as the sum of two terms. The first term is a force $-F$ that is positive if the phase into which the interface is moving has higher energy than the other phase; this force thus tries to locally stabilize the system. The second term $-bV$ represents a *drag force* opposing interfacial motion.

(2) By (6.4) and (6.5), the left side of (5.8) reduces to $-bV^2$. Thus tracing backwards the steps leading to (5.8), we see that the left side of the

dissipation inequality (5.3) minus the right is

$$-\int_\imath b(\theta, V)V^2\,ds, \tag{6.9}$$

so that $b(\theta, V)V^2$ represents the *total energy dissipated*, per unit length.

6.2 Balance of capillary forces revisited; corners

By (1.3) and (6.4), the *capillary force* (4.1) has the form

$$\mathbf{C} = \mathbf{C}(\theta) = f(\theta)\mathbf{T}(\theta) + f'(\theta)\mathbf{N}(\theta). \tag{6.10}$$

The relations (1.5) and (6.4) also imply that $\sigma_s = \xi K$ in every constitutive process. Tangential force balance $(4.4)_2$ then requires $B_{\tan} = 0$; granted this, we may use (6.5) to write the force balance $(4.2)_1$ in the equivalent integral form

$$\int_{\partial\imath}\mathbf{C}(\theta) = \int_\imath [F + b(\theta, V)V]\mathbf{N}(\theta)\,ds \tag{6.11}$$

for all evolving subcurves $\imath$ of $\jmath$, or, by $(4.4)_1$, (6.4), and (6.5), in the local form

$$b(\theta, V)V = [f(\theta) + f''(\theta)]K - F. \tag{6.12}$$

The relations (6.11) and (6.12) are equivalent for smooth evolving curves. For piecewise-smooth curves (6.12) must be augmented by additional conditions that ensure balance of capillary forces across corners defined by jump discontinuities in the dependence of $\theta(s, t)$ on s. Balance of forces requires the continuity of $\mathbf{C}(\theta(s, t))$ in s (cf. (4.5)); thus at a corner defined by a jump in $\theta(s, t)$ from θ^- to θ^+:

$$\mathbf{C}(\theta^-) = \mathbf{C}(\theta^+). \tag{6.13}$$

This result shows that *the set of corners consistent with balance of forces is a constitutive property of the material.*

Cusps, which are jumps in normal angle of amount π, are not possible: since $\mathbf{N}(\theta) = -\mathbf{N}(\theta + \pi)$ and $\mathbf{T}(\theta) = -\mathbf{T}(\theta + \pi)$, if we take the inner product of $\mathbf{T}(\theta)$ with $\mathbf{C}(\theta) = \mathbf{C}(\theta + \pi)$, we conclude, with the aid of (6.10), that $f(\theta) = -f(\theta + \pi)$, which contradicts the assumptions $f > 0$.

7
DIGRESSION: STATICAL THEORY OF INTERFACIAL STABILITY; CONVEXITY, THE FRANK DIAGRAM, AND CORNERS; WULFF REGIONS

It is a basic physical premise that stable configurations of a mechanical system minimize energy. In this section we will discuss the ramifications of this premise for the phase interface. Aside from being important in its own right, the discussion will furnish a derivation of the (statical) balance relations for capillary forces as Euler–Lagrange equations[20] corresponding to stationary values of the energy, and will provide a physical basis for certain assumptions we will make concerning the interfacial energy.

7.1 Preliminaries; polar diagrams

Throughout this section we will consider $\mathbf{N} = \mathbf{N}(\theta)$ and $\mathbf{T} = \mathbf{T}(\theta)$ as defined by (1.3). Then any $\mathbf{z} \in \mathbb{R}^2 \setminus \{\mathbf{0}\}$ admits the unique representation

$$\mathbf{z} = |\mathbf{z}|N(\theta);$$

θ, so defined, will be referred to as the **angle** of $\mathbf{z}$.

Let $p(\theta) > 0$ be a 2π-periodic function on $\mathbb{R}$. The **polar diagram** $P = \text{Polar}(p)$ of $p(\theta)$ is the graph, in polar coordinates (r, θ), of the function $r = p(\theta)$; points $(p(\theta), \theta)$ on P will usually be identified with the angle θ. Let ℓ be a line tangent to P, and let $\mathscr{T}$ denote the corresponding set of **tangency angles** (angles θ with $(p(\theta), \theta)$ points of tangency). If $\mathscr{T}$ contains two or more angles, then ℓ is a multitangent; if $\mathscr{T}$ contains exactly two angles, then ℓ is a bitangent.

There will be an ambiguity in our use of the term **convex**: in discussing curves, 'convex' means[21] 'strictly convex'; for functions, polar diagrams, and regions, 'convex' means 'convex but not necessarily strictly convex'.

A polar diagram P is **globally convex** at an angle θ if, letting $\mathbf{z}$ denote the point of P with angle θ,

$$\mathbf{M} \cdot (\mathbf{z} - \mathbf{y}) \geq 0 \tag{7.1}$$

---
[20] In the materials science literature, interfacial balance laws are usually derived as Euler–Lagrange equations corresponding to stationary values of a global Gibbs function. While such variational derivations often point the way toward a correct statement of the relevant law, they tend to obscure the fundamental nature of force-balance laws as *basic axioms* in any general dynamical framework that includes dissipation.
[21] In fact, it means more, since the curvature cannot vanish even at isolated points.

whenever $\mathbf{y} \in P$ and $\mathbf{M}$ is an outward normal to P at $\mathbf{z}$ (outward from the bounded region enclosed by P). This global convexity is **strict** if the inequality is strict for $\mathbf{y} \neq \mathbf{z}$. **Local convexity** and **local strict convexity** are defined analogously, with $\mathbf{y}$ restricted to a sufficiently small neighbourhood of $\mathbf{z}$. The *connected components* of the set of globally convex angles will be referred to as **globally convex sections** of P. If (α, β) is the open interval that separates two *adjacent* globally convex sections, then α and β are the tangency angles of a multitangent. We will refer to such multitangents as **convexifying multitangents**.

The **convexification** Convex(P) of P is the boundary of the convex hull of P. An angle is then globally convex if and only if it belongs to the intersection of Convex(P) with P. Further, the flat sections on which Convex(P) does not coincide with P lie on convexifying multitangents.

7.2 Convexity; the extended and convexified energies, and the Frank diagram

Important to discussions of interfacial stability is the convexity of the interfacial energy $f(\theta)$. This convexity is most easily phrased in terms of the **extended energy** $\bar{f}$, which is the *homogeneous extension* of f to $\mathbb{R}^2 \setminus \{\mathbf{0}\}$:

$$\bar{f}(\mathbf{z}) = |\mathbf{z}| f(\theta) \tag{7.2}$$

for $\mathbf{z} \in \mathbb{R}^2 \setminus \{\mathbf{0}\}$, where θ is the angle of $\mathbf{z}$. The convexity of $\bar{f}$ is related to the convexity of its level sets and, what is most important,

the level set $\{\mathbf{z}: \bar{f}(\mathbf{z}) = 1\}$ is the polar diagram of $f(\theta)^{-1}$. (7.3)

We will refer to this polar diagram as the **Frank diagram**.[22]

(7A) Properties of the extended energy

(a) *Let $\mathbf{y}$ be a non-zero vector with angle θ, and let $\lambda > 0$. Then*

$$\bar{f}(\lambda \mathbf{y}) = \lambda \bar{f}(\mathbf{y}), \qquad \bar{f}(\mathbf{y}) = \mathrm{grad}\, \bar{f}(\mathbf{y}) \cdot \mathbf{y},$$
$$\mathrm{grad}\, \bar{f}(\lambda \mathbf{y}) = \mathrm{grad}\, \bar{f}(\mathbf{y}), \qquad \mathrm{grad}\, \bar{f}(\mathbf{y}) = f(\theta) \mathbf{N}(\theta) - f'(\theta) \mathbf{T}(\theta). \tag{7.4}$$

(b) *Let $\mathscr{T}$ denote the tangent to the Frank diagram at a point $\mathbf{z}$. Then $\mathrm{grad}\, \bar{f}(\mathbf{z})$ is outward normal to the Frank diagram at $\mathbf{z}$, while $|\mathrm{grad}\, \bar{f}(\mathbf{z})|^{-1}$ is the perpendicular distance from the origin to $\mathscr{T}$.*

(c) *For $\mathbf{y}, \mathbf{z} \in \mathbb{R}^2 \setminus \{\mathbf{0}\}$, $\mathrm{grad}\, \bar{f}(\mathbf{z}) = \mathrm{grad}\, \bar{f}(\mathbf{y})$ if and only if the angles of $\mathbf{y}$ and $\mathbf{z}$ are tangency angles of a multitangent to the Frank diagram.*

[22] Frank (1963).

STATICAL THEORY OF INTERFACIAL STABILITY 43

(d) *If* $\mathbf{y}, \mathbf{z} \in \mathbb{R}^2 \setminus \{\mathbf{0}\}$ *and* $\omega > 0$ *is such that* $f(\omega \mathbf{z}) = f(\mathbf{y})$, *then*

$$f(\mathbf{y}) - f(\mathbf{z}) - \operatorname{grad} f(\mathbf{z}) \cdot (\mathbf{y} - \mathbf{z}) = \omega \operatorname{grad} f(\mathbf{z}) \cdot (\mathbf{z} - \omega^{-1} \mathbf{y}). \quad (7.5)$$

PROOF.

(a) The definition (7.2) implies (7.4)$_1$, and this, in turn, implies (7.4)$_{2,3}$. The gradient, in polar coordinates (r, θ), of a function $g(\mathbf{z})$ is given by

$$\operatorname{grad} g = g_r \mathbf{N} - r^{-1} g_\theta \mathbf{T}$$

and (7.4)$_4$ follows from (7.2).

(b) The outward normality of $\operatorname{grad} f(\mathbf{z})$ to the Frank diagram follows from (7.3) and the fact that f increases with the radial distance from the origin. The point $\mathbf{z}$ on the Frank diagram has the representation $\mathbf{z} = f(\theta)^{-1} \mathbf{N}(\theta)$, with θ the angle of $\mathbf{z}$, and $\mathbf{M} = \operatorname{grad} f(\mathbf{z}) / |\operatorname{grad} f(\mathbf{z})|$ is normal to $\mathcal{T}$ and directed away from the origin. Thus $\mathbf{M} \cdot [f(\theta)^{-1} \mathbf{N}(\theta)]$ is the perpendicular distance from the origin to $\mathcal{T}$; by (7.4)$_4$ this distance is $|\operatorname{grad} f(\mathbf{z})|^{-1}$.

(c) By (7.4)$_3$, we may restrict attention to points $\mathbf{y}, \mathbf{z}$ on the Frank diagram. Let $\mathcal{T}(\mathbf{y})$ and $\mathcal{T}(\mathbf{z})$ denote the tangents to the Frank diagram at $\mathbf{y}$ and $\mathbf{z}$. If $\operatorname{grad} f(\mathbf{z}) = \operatorname{grad} f(\mathbf{y})$, then, by (b), $\mathcal{T}(\mathbf{y})$ and $\mathcal{T}(\mathbf{z})$ are parallel and equidistant from the origin. Thus either $\mathcal{T}(\mathbf{y}) = \mathcal{T}(\mathbf{z})$ or the angles of $\mathbf{y}$ and $\mathbf{z}$ differ by π; the latter alternative contradicts $\operatorname{grad} f(\mathbf{z}) = \operatorname{grad} f(\mathbf{y})$. On the other hand, if $\mathcal{T}(\mathbf{y}) = \mathcal{T}(\mathbf{z})$, then, by (b), $\operatorname{grad} f(\mathbf{z}) = \operatorname{grad} f(\mathbf{y})$.

(d) By (7.4)$_{2,3}$,

$$f(\mathbf{y}) - f(\mathbf{z}) = f(\omega \mathbf{z}) - f(\mathbf{z}) = \operatorname{grad} f(\mathbf{z}) \cdot (\omega \mathbf{z} - \mathbf{z}),$$

which implies (7.5). ■

Let $\mathbf{z} \neq \mathbf{0}$. The function f is **globally convex** at $\mathbf{z}$, if

$$f(\mathbf{y}) - f(\mathbf{z}) \geq \operatorname{grad} f(\mathbf{z}) \cdot (\mathbf{y} - \mathbf{z}) \quad (7.6)$$

for all $\mathbf{y} \neq \mathbf{0}$, which is equivalent to the requirement that

$$f(\mathbf{z}) \leq a f(\mathbf{p}) + (1 - a) f(\mathbf{q}) \quad (7.7)$$

whenever $\mathbf{z} = a\mathbf{p} + (1 - a)\mathbf{q}$ with $\mathbf{p}, \mathbf{q} \neq \mathbf{0}$, $a \in (0, 1)$; f is **convex** if it is globally convex at every non-zero vector. (To show that (7.6) implies (7.7) add a times (7.6) with $\mathbf{y} = \mathbf{p}$ to $(1 - a)$ times (7.6) with $\mathbf{y} = \mathbf{q}$. To establish the converse assertion fix $\mathbf{h} \in \mathbb{R}^2$ and take $\mathbf{q} = \mathbf{z} - a\mathbf{h}$, $\mathbf{p} = \mathbf{z} + (1 - a)\mathbf{h}$, and $\varphi(a) = a f(\mathbf{p}) + (1 - a) f(\mathbf{q}) - f(\mathbf{z})$; then $\varphi(0) = 0$ and (7.7) yields $\varphi(a) \geq 0$ for $a > 0$ small, so that $\varphi'(0) \geq 0$, which implies (7.6).)

Since f is homogeneous, it cannot be strictly convex in the usual sense. We say that f is **globally strictly convex at z as a homogeneous function** if (7.6) holds whenever $\mathbf{y} \neq \mathbf{0}$ and the angle of $\mathbf{y}$ differs from that of $\mathbf{z}$. **Local convexity** and **local strict convexity as a homogeneous function** are defined analogously.

(7B) Convexity theorem *Assertions* (a) *and* (b) *are equivalent*:

(a) *the Frank diagram is globally convex at* θ;

(b) f *is globally convex at any* $z \neq 0$ *with angle* θ.

Moreover, (a) *and* (b) *remain equivalent when 'globally' is replaced by 'locally', 'globally strictly', or 'locally strictly', provided 'strictly' in* (b) *is as a homogeneous function.*

PROOF. Let z be a point of the Frank diagram. By (b) of (7A), we may replace the inequality (7.1) by

$$\operatorname{grad} f(z) \cdot (z - y) \geq 0 \qquad (7.8)$$

for y on the Frank diagram. Further, for z on the Frank diagram and $y \in \mathbb{R}^2 \setminus \{0\}$, there is an $\omega > 0$ such that $\omega^{-1} y$ lies on the Frank diagram. For this ω, $f(\omega z) = f(y)$, so that, by (7.5),

$$f(y) - f(z) \geq \operatorname{grad} f(z) \cdot (y - z) \Leftrightarrow \operatorname{grad} f(z) \cdot (z - \omega^{-1} y) \geq 0. \quad (7.9)$$

The last two results imply the equivalence of (a) and (b), and of (a) and (b) with 'globally' replaced by 'locally'. The remaining assertions are established analogously. ∎

Another condition equivalent to the global convexity of the Frank diagram at θ is that:

$$\operatorname{grad} f(N(\theta)) \cdot N(\alpha) \leq f(N(\alpha)) \qquad (7.10)$$

for all angles α. To verify (7.10), let z and y have angles θ and α, respectively. Then, using (7.2) and (7.4), we see that (7.10) is equivalent to $\operatorname{grad} f(z) \cdot y \leq f(y)$, and hence to

$$\operatorname{grad} f(z) \cdot (y - z) \leq f(y) - f(z);$$

thus (7.10) holds for all α if and only if f is globally convex at z.

The Frank diagram is the locus of the **Frank potential**

$$\sigma(\theta) = f(\theta)^{-1} N(\theta),$$

and, by (6.10), generates the capillary force through the relation

$$C(\theta) = -f(\theta)^2 \sigma'(\theta). \qquad (7.11)$$

This is not the only relation between the capillary force and the Frank diagram.

(7C) Relations between the capillary force, the Frank diagram, and the extended energy

(a) *The capillary force at any θ is given by*[23]

$$\mathbf{C}(\theta) = \mathbf{Q}\operatorname{grad} f(\mathbf{y}) \tag{7.12}$$

with $\mathbf{y}$ any non-zero vector with angle θ and $\mathbf{Q}$ the orthogonal transformation that rotates vectors clockwise by $\pi/2$.

(b) *$\mathbf{C}(\theta)$ is tangent to the Frank diagram and points in the direction of decreasing θ, and $|\mathbf{C}(\theta)|^{-1} = |\operatorname{grad} f(\mathbf{y})|^{-1}$ is the perpendicular distance from the origin to the Frank diagram's tangent at θ.*

(c) *$\mathbf{C}(\theta_1) = \mathbf{C}(\theta_2)$ if and only if θ_1 and θ_2 are tangency angles of a multitangent to the Frank diagram.*

PROOF. The relation (7.12) follows from (6.10) and (7.4)$_4$; (b) and (c) follow from (7.12) in conjunction with (b) and (c) of Theorem 7.1. ∎

In the next lemma α and β are angles with $0 < \beta - \alpha < \pi$, and $\mu_\alpha(\theta)$ and $\mu_\beta(\theta)$ are the facet densities of α-facets and β-facets as defined by (1.19).

(7D) Lemma *The following are equivalent:*

(a) *The Frank diagram is a straight line between α and β.*

(b) *$\mathbf{C}(\theta)$ is constant for $\theta \in [\alpha, \beta]$.*

(c) *$f(\theta) = \mu_\alpha(\theta) f(\alpha) + \mu_\beta(\theta) f(\beta)$ for $\theta \in [\alpha, \beta]$.*

(d) *$f(\theta) + f''(\theta) = 0$ for $\theta \in [\alpha, \beta]$.*

PROOF. The equivalence of (a) and (b) follows from (b) of (7C). Further, by (1.4) and (6.10),

$$\mathbf{C}'(\theta) = [f(\theta) + f''(\theta)]\mathbf{N}(\theta), \tag{7.13}$$

so that (b) and (d) are equivalent, and, in view of (1.21), (c) implies (d).

To complete the proof, we have only to show that (b) implies (c). Assume that (b) holds. Then $\mathbf{C}(\theta) = \mathbf{C}(\alpha) = \mathbf{C}(\beta)$, so that, by (6.10) and (1.19),

$$f(\theta) = \mathbf{C}(\theta) \cdot \mathbf{T}(\theta) = \mathbf{C}(\theta) \cdot [\mu_\alpha(\theta)\mathbf{T}(\alpha) + \mu_\beta(\theta)\mathbf{T}(\beta)] = \mu_\alpha(\theta) f(\alpha) + \mu_\beta(\theta) f(\beta),$$

which completes the proof. ∎

---

[23] The vector field $\operatorname{grad} f$ is the ξ-vector of Cahn and Hoffman (1972, 1974).

The convexified Frank diagram is the Frank diagram of a modified energy $f^\#(\theta)$, called the **convexified energy**. This energy satisfies: $f^\#(\theta) = f(\theta)$ whenever θ is a globally convex angle of the Frank diagram, $f^\#(\theta) < f(\theta)$ otherwise. We will also use the **extended convexified energy** $f^\#(\mathbf{z})$: $f^\#(\mathbf{z})$ is the homogeneous extension of $f^\#$ to $\mathbb{R}^2\setminus\{\mathbf{0}\}$. By (7B), $f^\#$ is a convex function (in fact, $f^\#$ is the convexification of f).

(7E) Properties of the convexified energy *Let α and β, with $0 < \beta - \alpha < \pi$, be tangency angles of a convexifying bitangent to the Frank diagram, and let $\mu_\alpha(\theta)$ and $\mu_\beta(\theta)$ denote the facet densities of α-facets and β-facets as defined by (1.19). Then, for $\alpha \le \theta \le \beta$,*

$$f^\#(\theta) = \mu_\alpha(\theta)f(\alpha) + \mu_\beta(\theta)f(\beta). \tag{7.14}$$

PROOF. The relation (7.14) follows from (7D), since $f(\alpha) = f^\#(\alpha)$ and $f(\beta) = f^\#(\beta)$, and since the convexified Frank diagram is a straight line between α and β. ∎

Exercise Let ω be an α, β wrinkled curve, let $\imath_{\alpha\beta}$ be an infinitesimally wrinkled curve with normal angles α and β, let ℓ be an oriented line segment, and assume that the initial points of ω, ℓ, and $\imath$ coincide, and similarly for their terminal points. Show that

$$\int_\omega f(\theta)\,ds = \int_\ell f^\#(\theta)\,ds = \int_{\imath_{\alpha\beta}} f(\theta)\,ds = \int_\imath f^\#(\theta)\,ds. \tag{7.15}$$

The next lemma gives an inequality that is useful in comparing the energy of curves.

(7F) Lemma *Let $\lambda > 0$, $\theta \ne \theta_0$. If θ_0 is a globally convex angle of the Frank diagram, then*

$$f(\theta) - \lambda f(\theta_0) \ge \operatorname{grad} f(\mathbf{N}(\theta_0)) \cdot [\mathbf{N}(\theta) - \lambda \mathbf{N}(\theta_0)] \tag{7.16}$$

with equality if and only if θ and θ_0 are tangency angles for a multitangent to the Frank diagram.

PROOF. Write $\mathbf{N} = \mathbf{N}(\theta)$, $\mathbf{N}_0 = \mathbf{N}(\theta_0)$. The inequality (7.16) follows from (7B) and (7.6) with $\mathbf{y} = \mathbf{N}$, $\mathbf{z} = \lambda \mathbf{N}_0$.

Choose $\tau > 0$ such that $\mathbf{z} = \tau \mathbf{N}_0$ lies on the Frank diagram; let $\mathbf{y} = \tau\lambda^{-1}\mathbf{N}$ and choose $\omega > 0$ such that $\omega^{-1}\mathbf{y}$ also lies on the Frank diagram (and hence $f(\omega\mathbf{z}) = f(\mathbf{y})$). Then (7.5) and the homogeneity of f imply that

$$f(\mathbf{N}) - \lambda f(\mathbf{N}_0) - \operatorname{grad} f(\mathbf{N}_0)\cdot[\mathbf{N} - \lambda\mathbf{N}_0] = \frac{\lambda\omega}{\tau}\operatorname{grad} f(\mathbf{z})\cdot(\mathbf{z} - \omega^{-1}\mathbf{y}). \tag{7.17}$$

Since θ_0 is a globally convex angle, we may conclude from (7.8) (with **y** replaced by $\omega^{-1}\mathbf{y}$) that the right side of (7.17) is non-negative. Further, since grad $f(\mathbf{z})$ is normal to the Frank diagram at **z**, the right side of (7.17) vanishes if and only if θ (the angle of **z**) and θ_0 (the angle of $\omega^{-1}\mathbf{y}$) are tangency angles of a multitangent to the Frank diagram. ∎

7.3 Stability

It is convenient to write

$$\mathscr{F}(\imath) = \int_{\imath} f(\theta)\, ds \qquad (7.18)$$

for the energy of a PS curve $\imath$, and to refer to $\imath$ as **stable** if

$$\mathscr{F}(\imath) \leq \mathscr{F}(c) \qquad (7.19)$$

for all PS curves c whose initial and terminal points coincide with those of $\imath$. The next theorem gives the complete list of stable curves. For convenience, we use the term 'multitangent' as shorthand for 'multitangent to the Frank diagram', and similarly for the term 'bitangent'.

(7G) Stability theorem[24] *A PS curve $\imath$ is stable if and only if either*:

(a) *$\imath$ is a facet whose normal angle is a globally convex angle of the Frank diagram, or*
(b) *$\imath$ is a PS curve whose angle-set contains only tangency angles of a convexifying multitangent $\mathscr{M}$ (so that $\imath$ is fully faceted if the set of tangency angles of $\mathscr{M}$ is discrete; $\imath$ is a wrinkled curve if $\mathscr{M}$ is a bitangent).*

PROOF. Let

$$\mathscr{F}^{\#}(c) = \int_{c} f^{\#}(\theta)\, ds \qquad (7.20)$$

denote the energy a PS curve c would have were $f^{\#}$ the interfacial energy. Further, let ℓ denote the facet whose initial and terminal points coincide with those of $\imath$, let L denote the length and θ_0 the normal angle of ℓ, and let

$$\mathbf{N}_0 = \mathbf{N}(\theta_0), \qquad \mathbf{T}_0 = \mathbf{T}(\theta_0), \qquad \lambda(\theta) = \mathbf{T}_0 \cdot \mathbf{T}(\theta).$$

Let c be a PS curve whose initial and terminal points coincide with those of ℓ. Then

$$\mathbf{N}(\theta) = \lambda(\theta)\mathbf{N}_0 + [\mathbf{T}_0 \cdot \mathbf{N}(\theta)]\mathbf{T}_0,$$
$$\int_c \lambda(\theta)\,ds = L, \qquad \int_c \mathbf{T}_0 \cdot \mathbf{N}(\theta)\, ds = 0, \qquad (7.21)$$

---

[24] The main ideas are due to Herring (1951b). See also Mullins and Sekerka (1962); Cahn and Hoffman (1974).

so that
$$\int_c [\mathbf{N}(\theta) - \lambda(\theta)\mathbf{N}_0] \, ds = \mathbf{0}. \tag{7.22}$$

Using $(7.21)_2$,

$$\mathscr{F}^{\#}(c) - \mathscr{F}^{\#}(\ell) = \int_c f^{\#}(\theta) \, ds - \int_\ell f^{\#}(\theta_0) \, ds$$

$$= \int_c [f^{\#}(\theta) - \lambda(\theta)f^{\#}(\theta_0)] \, ds, \tag{7.23}$$

and, by (7F) and (7.22) applied to the convexified energy $f^{\#}$, $\mathscr{F}^{\#}(\ell) \leq \mathscr{F}^{\#}(c)$ with equality for $c \neq \ell$ if and only if θ_0 and all of the normal angles of c are tangency angles of a multitangent to the convexified Frank diagram. This and the remarks of the paragraph preceding (7E) yield the result

$$\mathscr{F}^{\#}(\ell) \leq \mathscr{F}(c)$$

with equality if and only if $c = \ell$ and θ_0 is globally stable or $c \neq \ell$ and θ_0 and all of the normal angles of c are tangency angles of a multitangent to the Frank diagram. This yields the desired conclusions. ∎

The energy of an infinitesimally wrinkled curve $\imath_{\alpha\beta}$ is well defined by

$$\mathscr{F}(\imath_{\alpha\beta}) = \int_{\imath_{\alpha\beta}} f(\theta) \, ds$$

(cf. (1.25)), which allows us to extend the stability theorem to such curves. With this in mind, let us agree to use the term **generalized curve** for a 'curve' that is either a PS curve or an infinitesimally wrinkled curve, and to refer to a generalized curve $\imath$ as **stable** if $\mathscr{F}(\imath) \leq \mathscr{F}(c)$ for all generalized curves c whose initial and terminal points coincide with those of $\imath$. In view of (1E), the energy of an infinitesimal wrinkled curve $\imath_{\alpha\beta}$ coincides with that of any α, β wrinkled curve ω whose initial and terminal points coincide with those of $\imath_{\alpha\beta}$. We therefore have the following corollary of (7G).

(7H) Generalized stability theorem *A generalized curve $\imath$ is stable if and only if either $\imath$ is a PS curve of one of the two types specified in (a) and (b) of (7G), or $\imath$ is an infinitesimally wrinkled curve $\imath_{\alpha\beta}$ whose facets have normal angles α and β that are tangency angles of a convexifying multitangent.*

(7I) Remark Let $\mathbf{x}$ and $\mathbf{y}$ be given points, and let $\mathscr{P}$ denote the set of generalized curves from $\mathbf{x}$ to $\mathbf{y}$. Let θ_0 denote the normal angle of the oriented straight line (facet) $\ell \in \mathscr{P}$. We then have the following consequences of the stability theorem. If θ_0 is globally convex, then ℓ minimizes energy among all generalized curves in $\mathscr{P}$, but for θ_0 *not* globally convex there are wrinkled curves in $\mathscr{P}$ with lower energy. In fact, for such θ_0 there are adjacent tangency angles α and β of a convexifying multitangent $\mathscr{M}$ to the Frank diagram such that $0 < \beta - \alpha < \pi$ and $\theta_0 \in (\alpha, \beta)$; and any wrinkled curve or infinitesimally wrinkled curve $\omega \in \mathscr{P}$ with α and β as normal angles is a minimizer over $\mathscr{P}$, with the energy of ω strictly less than that of ℓ, or of any other curve in $\mathscr{P}$ whose angle-set contains angles that are not tangency angles of $\mathscr{M}$. This demonstrates the importance of allowing infinitesimally wrinkled curves:

A PS curve with angle-set in $[\alpha, \beta]$ (that is not an α, β wrinkled curve) can lower its energy by developing appropriate infinitesimal wrinkles, even through it does not otherwise alter its shape. (7.24)

Local minima of the interfacial energy are also of interest. Let ℓ be a facet with normal angle θ_0. Then ℓ is **locally stable** if there is an $\varepsilon > 0$ such that $\mathscr{F}(\imath) \geq \mathscr{F}(\ell)$ for any PS curve $\imath$ whose initial and terminal points coincide with those of ℓ and whose normal angle $\theta(s)$ satisfies $|\theta(s) - \theta_0| < \varepsilon$ for all s.

(7J) Local stability theorem[25] *The following are equivalent*:

(a) *the Frank diagram is locally convex at θ_0*;

(b) *any facet with normal angle θ_0 is locally stable.*

Moreover, each of these implies $f(\theta_0) + f''(\theta_0) \geq 0$ and is implied by $f(\theta_0) + f''(\theta_0) > 0$.

PROOF. Throughout the proof:

$$\mathbf{N}_0 = \mathbf{N}(\theta_0), \qquad \mathbf{T}_0 = \mathbf{T}(\theta_0).$$

The argument in the paragraph containing (7.23) here yields the assertion (a) $\Rightarrow$ (b).

Assume that (b) holds. Then, in particular, $\mathscr{F}(\omega) \geq \mathscr{F}(\ell)$ for any α, β wrinkled curve ω whose initial and terminal points coincide with ℓ, provided the normal angles α and β of its facets are sufficiently close to θ_0. Thus, writing L for the length of ℓ, and L_α and L_β for the total lengths of α-facets and β-facets, and using (1.17) and (1.18), we conclude that

$$f(\mathbf{N}_0) \leq a_1 f(\mathbf{N}_1) + a_2 f(\mathbf{N}_2)$$

whenever $\mathbf{N}_0 = a_1 \mathbf{N}_1 + a_2 \mathbf{N}_2$ with $a_1, a_2 > 0$ and $\mathbf{N}_1$ and $\mathbf{N}_2$ unit vectors sufficiently close to $\mathbf{N}_0$. Since f is homogeneous, this implies that (7.7) is satisfied with $\mathbf{z} = \mathbf{N}_0$ whenever $\mathbf{N}_0 = a\mathbf{p} + (1-a)\mathbf{q}$ with $a \in (0, 1)$ and $\mathbf{p}, \mathbf{q} \neq 0$ close to $\mathbf{N}_0$. Thus $\mathbf{N}_0$ is a point of local convexity of f, so that, by the convexity theorem (7B), (a) is satisfied. Hence (a) and (b) are equivalent.

---

[25] Cf. Herring (1951*b*); Frank (1963); Gjostein (1963); Gruber as referred to in Gjostein (1963); Taylor (1978); Fonsaca (1989).

We show next that (a) implies $f(\theta_0) + f''(\theta_0) \geq 0$. Assume that (a) holds. Then, by the convexity theorem and (7.6), the function (cf. (7.4)$_4$)

$$\varphi(\theta) = f(\mathbf{N}(\theta)) - f(\mathbf{N}_0) - \text{grad } f(\mathbf{N}_0) \cdot (\mathbf{N}(\theta) - \mathbf{N}_0)$$
$$= f(\theta) - f(\theta_0) - [f(\theta_0)\mathbf{N}_0 - f'(\theta_0)\mathbf{T}_0] \cdot (\mathbf{N}(\theta) - \mathbf{N}_0)$$

has a local minimum at θ_0, so that $\varphi''(\theta_0) \geq 0$. But, by (1.4), $\varphi''(\theta_0) = f(\theta_0) + f''(\theta_0)$. Thus $f(\theta_0) + f''(\theta_0) \geq 0$.

Our final step will be to show that $f(\theta_0) + f''(\theta_0) > 0$ implies (a). Let

$$\mathbf{y}(\theta) = f(\theta)^{-1}\mathbf{N}(\theta), \qquad \mathbf{y}_0 = \mathbf{y}(\theta_0), \tag{7.25}$$

so that $\mathbf{y}(\theta)$ is the point on the Frank diagram with angle θ. In view of the definition containing (7.1), to verify (a) we must show that

$$\omega(\theta) = \text{grad } f(\mathbf{y}_0) \cdot (\mathbf{y}(\theta) - \mathbf{y}_0) \leq 0 \tag{7.26}$$

for all θ near θ_0. Since grad $f(\mathbf{y}_0)$ and $\mathbf{y}'(\theta)$ are orthogonal, $\omega'(\theta_0) = 0$. Thus, since $\omega(\theta_0) = 0$, it suffices to show that

$$\omega''(\theta_0) < 0.$$

By (1.4) and (7.25),

$$f^2\mathbf{y}'' = -[f + f'' - 2f^{-1}(f')^2]\mathbf{N} + 2f'\mathbf{T}.$$

Thus, in view of (7.4)$_4$,

$$\omega''(\theta_0) = [f(\theta_0)\mathbf{N}_0 - f'(\theta_0)\mathbf{T}_0] \cdot \mathbf{y}''(\theta_0) = -f(\theta_0)^{-1}[f(\theta_0) + f''(\theta_0)],$$

so that $f(\theta_0) + f''(\theta_0) > 0$ implies (a). ∎

The Frank diagram is flat (a straight line) between the angles θ_1 and θ_2 if and only if $\omega(\theta, \theta_0)$ defined by (7.26) vanishes at all $\theta, \theta_0 \in [\theta_1, \theta_2]$, or equivalently, $\omega_{\theta_0}(\theta, \theta_0) = 0$ at all such θ, θ_0. Since

$$\omega_{\theta_0}(\theta, \theta_0) = -\frac{[f(\theta_0) + f''(\theta_0)]\mathbf{T}_0 \cdot \mathbf{N}(\theta)}{f(\theta)},$$

we have the following result:

> The Frank diagram is flat between the angles θ_1 and θ_2 if and only if $f(\theta) + f''(\theta) = 0$ for all $\theta \in [\theta_1, \theta_2]$. (7.27)

The following terminology is guided by the global and local stability theorems. We refer to the globally convex angles and the globally convex sections of the Frank diagram as **globally stable angles** and **globally stable sections** of the interfacial energy f. We say that f is **strictly stable at** θ, **stable at** θ, or **unstable at** θ according as

$$f(\theta) + f''(\theta) > 0, \qquad f(\theta) + f''(\theta) \geq 0, \qquad f(\theta) + f''(\theta) < 0; \tag{7.28}$$

and that f is: **strictly stable** if it is strictly stable for all θ; and **stable** if it is stable for all θ. Since $f(\theta) > 0$, the interfacial energy cannot be unstable for all θ.

(7K) Theorem *If $\mathbf{C}(\theta^-) = \mathbf{C}(\theta^+)$ with $0 < \theta^+ - \theta^- < \pi$, then either f is unstable somewhere in (θ^-, θ^+) or the Frank diagram is flat between θ^- and θ^+.*

PROOF. Choose $\mathbf{e}$ such that $\varphi(\theta) = \mathbf{e} \cdot \mathbf{N}(\theta) > 0$ on (θ^-, θ^+). Then, since $\mathbf{C}(\theta^-) = \mathbf{C}(\theta^+)$, if we integrate $\mathbf{e} \cdot \mathbf{C}'(\theta)$ from $\theta = \theta^-$ to $\theta = \theta^+$ we find, using (7.13), that

$$\int_{\theta^-}^{\theta^+} [f(\theta) + f''(\theta)] \varphi(\theta) \, d\theta = 0; \tag{7.29}$$

this and (7D) imply the desired conclusion. ∎

Theorem (7K) and the remark made in the last paragraph of Section 6.2 imply that

$$\begin{aligned} &\text{there are no solutions of } \mathbf{C}(\theta^-) = \mathbf{C}(\theta^+), \\ &\theta^- \neq \theta^+, \text{ if } f \text{ is strictly stable.} \end{aligned} \tag{7.30}$$

In view of the paragraph containing (6.13), an interface with corners is not possible when the interfacial energy is strictly stable: *corners require instability*.

To avoid technical difficulties, we will assume, for the remainder of Chapter 7, that the interfacial energy is **regular** in the sense of the following hypotheses.

(R1) The Frank diagram has no multitangents other than bitangents, and at most a finite number of bitangents.

(R2) All globally stable angles are strictly stable.

A property of regular f is that:

$$\text{Multitangency angles are isolated,} \tag{7.31}$$

for otherwise they would not be finite in number. (By a multitangency angle we mean a tangency angle of some multitangent.)

7.4 Instability of the total energy

The discussion thus far has concerned the interface without regard to the bulk material. We now focus attention on the interaction between the bulk and interfacial energies by considering the **total energy**

$$\mathcal{E}(\Omega) = F \text{ area}(\Omega) + \mathscr{F}(\partial\Omega), \qquad \mathscr{F}(\partial\Omega) = \int_{\partial\Omega} f(\theta) \, ds \qquad (7.32)$$

over the class of bounded regions Ω. In writing (7.32) it is tacit that Ω is occupied by phase 1 material with $F = \psi_1 - \psi_2$ (=constant) the bulk energy of phase 1 relative to phase 2 (cf. (5.10)). Further, when we refer to a **bounded region** Ω it will be tacit that $\partial\Omega$ is a closed PS curve with Ω as reference region, so that *the normal to $\partial\Omega$, considered as a PS curve, is the outward normal to Ω.*

The following definitions will be useful. Let Ω be a bounded region. A **variation** of Ω is a one-parameter family $\Gamma(t)$, $-T < t < T$, of bounded regions with $\Gamma(0) = \Omega$ and $\partial\Gamma(t)$, $-T < t < T$, a PS evolving curve. Ω is an **equilibrium** of the total energy $\mathcal{E}$ if, given any variation $\Gamma(t)$ of Ω, $\mathcal{E}(t) = \mathcal{E}(\Gamma(t))$ satisfies

$$\mathcal{E}^{\cdot}(0) = 0;$$

$\mathcal{E}$ is **unstable to a variation** $\Gamma(t)$ of Ω if $\mathcal{E}(t) = \mathcal{E}(\Gamma(t))$ satisfies

$$\mathcal{E}^{\cdot}(0) \neq 0 \quad \text{or}$$
$$\mathcal{E}^{\cdot}(0) = 0 \quad \text{and} \quad \mathcal{E}^{\cdot\cdot}(0) < 0.$$

$\mathcal{E}$ is **unstable** if, given any bounded region Ω, $\mathcal{E}$ is unstable to some variation of Ω.

We will consider a particular class of variations: those that are dilations of Ω. Precisely, the **dilation of Ω of amount** $a > 0$ is the set

$$a\Omega = \{a\mathbf{x} : \mathbf{x} \in \Omega\}.$$

We then have the *scaling rules*:

$$\text{area}(a\Omega) = a^2 \text{ area}(\Omega), \qquad \mathscr{F}(\partial(a\Omega)) = a\mathscr{F}(\partial\Omega). \qquad (7.33)$$

Further, given any $\lambda \neq 0$ and $T > 0$, the family $\Gamma(t) = (1 + \lambda t)\Omega$, $-T < t < T$, defines a variation of Ω; such variations are referred to as **variational dilations of Ω**.

(7L) Instability theorem *The total energy is unstable: given any bounded region Ω, the total energy is unstable to all variational dilations of Ω.*

PROOF. Consider a variational dilation $\Gamma(t) = (1 + \lambda t)\Omega$ of Ω and write $\mathcal{E}(t) = \mathcal{E}(\Gamma(t))$. Then, by (7.33),

$$\mathcal{E}(t) = (1 + \lambda t)^2 F \text{ area}(\Omega) + (1 + \lambda t)\mathscr{F}(\partial\Omega),$$

and

$$\mathcal{E}^{\cdot}(0) = 2\lambda F \text{ area}(\Omega) + \lambda \mathscr{F}(\partial\Omega), \qquad \mathcal{E}^{\cdot\cdot}(0) = 2\lambda^2 F \text{ area}(\Omega).$$

Assume that $\mathscr{E}\,'(0) = 0$. Then

$$2F\,\text{area}(\Omega) + \mathscr{F}(\partial\Omega) = 0, \tag{7.34}$$

so that $F < 0$, which yields $\mathscr{E}\,''(0) < 0$. ∎

(7M) Remark By the instability theorem, the total energy cannot have a local minimum within the class of bounded regions, at least relative to any topology that renders dilational variations $\Gamma(t)$ continuous in t. We can say more about this instability. Consider the two cases $F \geq 0$ and $F < 0$ and recall that $F = \psi_1 - \psi_2$ and that the argument Ω in $\mathscr{E}(\Omega)$ is the region occupied by phase 1.

Case 1. $F \geq 0$, so that phase 2 is the more stable phase. In this case $\mathscr{E}(\Omega) > 0$ for all bounded regions Ω, but the infimum of $\mathscr{E}$ is zero. Thus minimizing sequences for $\mathscr{E}$ necessarily correspond, in the limit, to vanishing area, indicating that the most stable configuration has all of $\mathbb{R}^2$ of phase 2.

Case 2. $F < 0$, so that phase 1 is the more stable phase. Here the infimum of $\mathscr{E}(\Omega)$ over all bounded regions Ω is $-\infty$, and minimizing sequences correspond, in the limit, to infinite area, indicating that the most stable configuration has all of $\mathbb{R}^2$ of phase 1. But there is also a 'local minimum': $|F|\,\text{area}(\Omega) < \mathscr{F}(\partial\Omega)$ for all Ω of sufficiently small area, so that $\mathscr{E}(\Omega) > 0$ for all such Ω; hence 'a region of zero area and zero perimeter' minimizes $\mathscr{E}(\Omega)$ over all Ω of small area. One therefore expects that a seed of sufficiently large area is needed to nucleate a phase 1 region. A dynamic rationale for this expectation will be given in Section 10 (cf. 10C).

Regions Ω that satisfy (7.34) are stationary with respect to dilations, but not necessarily with respect to arbitrary variations. The next theorem gives conditions that are necessary and sufficient for an equilibrium of the total energy.

(7N) Euler–Lagrange equations *Ω is an equilibrium of the total energy if and only if $\partial\Omega$ is consistent with balance of capillary forces in the sense that*

$$\int_{\partial \imath} \mathbf{C}(\theta) + \int_{\imath} F\mathbf{N}(\theta)\,ds = \mathbf{0} \tag{7.35}$$

on each subcurve $\imath$ of $\partial\Omega$, or equivalently,[26]

$$\begin{aligned}[f(\theta) + f''(\theta)]K &= F &&\text{on each smooth arc of } \partial\Omega, \\ \mathbf{C}(\theta_i^-) &= \mathbf{C}(\theta_i^+) &&\text{at each juncture } i \text{ of } \partial\Omega.\end{aligned} \tag{7.36}$$

We will refer to (7.36) *as the* **equilibrium equations**.

---

[26] Cf. Herring (1951a); Gjostein (1963); Cahn and Hoffman (1974).

(7O) Lemma *Let $\varphi(s)$ be a piecewise-continuous function on $\partial\Omega$, and let $\boldsymbol{\varphi}_i$ be a vector at each junction i. Suppose that*

$$\int_{\partial\Omega} \varphi(s)V(s,0)\,ds + \sum \boldsymbol{\varphi}_i \cdot \mathbf{R}_i'(0) = 0 \qquad (7.37)$$

for all variations $\Gamma(t)$ of Ω, where $V(s,t)$ is the normal velocity of $\partial\Gamma(t)$, $\mathbf{R}_i'(t)$ is the velocity of juncture i of $\partial\Gamma(t)$, and the sum is over the essential junctures i of $\partial\Omega$. Then

$$\varphi(s) \equiv 0, \qquad \boldsymbol{\varphi}_i = 0 \quad \text{for all } i. \qquad (7.38)$$

PROOF. Choose an arbitrary arc $\imath_0$ of $\partial\Omega$, and let $\imath(t)$ be the evolving curve established in (a) of (2K). Then (possibly by decreasing T) the region $\Gamma(t)$ obtained from Ω by replacing the boundary arc $\imath_0$ by $\imath(t)$ for $-T < t < T$ is a variation of Ω. For this variation (7.37) yields

$$\int_{S_1}^{S_2} \varphi(s)\omega(s)\,ds = 0,$$

with $[S_1, S_2]$ the interval of arc lengths of $\imath_0$. Since ω is an arbitrary smooth function on $[S_1, S_2]$ with compact support in (S_1, S_2), and $\imath_0$ is arbitrary, this implies $(7.38)_1$.

Next, choose an arbitrary junction j of $\partial\Omega$. Let $\imath_0$ and $\jmath_0$ be the arcs of $\partial\Omega$ that meet at j, with the terminal point of $\imath_0$ connected to the initial point of $\jmath_0$. By (b) of (2K) there are evolving curves $\imath(t)$ and $\jmath(t)$, $-T < t < T$, with $\imath(0) = \imath_0$ and $\jmath(0) = \jmath_0$, such that the terminal point of $\imath(t)$ and the initial point of $\jmath(t)$ move together with velocity $\mathbf{a}$, but neighbourhoods of the initial point of $\imath(t)$ and the terminal point of $\jmath(t)$ are motionless. Then (possibly by decreasing T) the region $\Gamma(t)$ obtained from Ω by replacing the boundary arcs $\imath_0$ and $\jmath_0$ by $\imath(t)$ and $\jmath(t)$ for $-T < t < T$ is a variation of Ω. By (7.37), $(7.38)_1$, and the fact that $\mathbf{a}$ and the junction j are arbitrary, this yields $(7.38)_2$. ∎

PROOF OF (7N). Let $\Gamma(t)$ be a variation of $\partial\Omega$. If we differentiate $\mathscr{F}(t) = \mathscr{F}(\partial\Gamma(t))$ using the transport theorem (2I), we find that

$$\mathscr{F}'(t) = \int_{\partial\Gamma(t)} [f'(\theta)\theta^\circ - f(\theta)KV]\,ds - \sum [f(\theta_i^+)v_i^+ - f(\theta_i^-)v_i^-], \qquad (7.39)$$

where here and in what follows we suppress the argument t when convenient. Next, $V_i^\pm = \mathbf{N}_i^\pm \cdot \mathbf{R}_i'$ and $v_i^\pm = \mathbf{T}_i^\pm \cdot \mathbf{R}_i'$ (cf. (2.14), (2.38)), so that (6.10) yields

$$[\mathbf{C}(\theta_i^+) - \mathbf{C}(\theta_i^-)] \cdot \mathbf{R}_i' = f'(\theta_i^+)V_i^+ + f(\theta_i^+)v_i^+ - [f'(\theta_i^-)V_i^- + f(\theta_i^-)v_i^-];$$

thus, since $\theta^\circ = V_s$ and $K = \theta_s$ (cf. (1.5), $(2.9)_2$), we conclude, with the aid of

(2.48), that

$$\int_{\partial\Gamma(t)} f'(\theta)\theta^\circ \, ds - \sum [f(\theta_i^+)v_i^+ - f(\theta_i^-)v_i^-]$$

$$= -\int_{\partial\Gamma(t)} f''(\theta)KV \, ds - \sum [\mathbf{C}(\theta_i^+) - \mathbf{C}(\theta_i^-)] \cdot \mathbf{R}_i^*.$$

Thus, using (2.49)$_1$ and noting that $\Gamma(0) = \Omega$,

$$\mathscr{E}^\bullet(0) = \int_{\partial\Omega} [F - g(\theta)K]V \, ds - \sum [\mathbf{C}(\theta_i^+) - \mathbf{C}(\theta_i^-)] \cdot \mathbf{R}_i^*, \quad (7.40)$$

where

$$g(\theta) = f(\theta) + f''(\theta), \quad (7.41)$$

and where all functions of t are evaluated at $t = 0$. Thus, appealing to Lemma 7O, we see that (7.39) vanishes for all variations of Ω if and only if (7.36) are satisfied. ∎

7.5 Equilibria of the total energy; Wulff regions

The set

$$\Lambda = \Lambda(f) = \{\mathbf{x} \in \mathbb{R}^2 \colon \mathbf{x} \cdot \mathbf{N}(\theta) \leq f(\theta) \text{ for all } \theta \in \mathbb{R}\} \quad (7.42)$$

is called the **Wulff region**[27] for f. This set is easily constructed: at each point **z** on the polar diagram of f draw a line through **z** perpendicular to **z**; the Wulff region is the set of points that can be reached from the origin without crossing any of these lines.

When we omit mention of f it will be understood that Λ is the Wulff region for f; we will, in fact, use the Wulff region $\Lambda(e)$ for functions $e(\theta) > 0$ other than the energy.

(7P) Properties of the Wulff region

(a) Λ is convex.

(b) *The angle-set of $\partial\Lambda$ is the set Θ of globally stable angles.*

(c) *$\partial\Lambda$ is the locus of the vector function*

$$\mathbf{r}_0(\theta) = f(\theta)\mathbf{N}(\theta) - f'(\theta)\mathbf{T}(\theta) \quad (7.43)$$

for $\theta \in \Theta$, and $\mathbf{r}_0(\theta)$ has the equivalent representations:

$$\begin{aligned}\mathbf{r}_0(\theta) &= \operatorname{grad} \bar{f}(\mathbf{y}), \\ \mathbf{r}_0(\theta) &= \mathbf{Q}\mathbf{C}(\theta),\end{aligned} \quad (7.44)$$

*where **y** is any non-zero vector with angle θ, while **Q** is the orthogonal transformation that rotates vectors counterclockwise by $\pi/2$.*

---

[27] Cf. Taylor (1978), who uses the term 'Wulff crystal'.

(d) *The curvature K_0 and support function p_0 of $\partial \Lambda$ are given by*

$$K_0(\theta) = -[f(\theta) + f''(\theta)]^{-1}, \qquad p_0(\theta) = f(\theta) \qquad (7.45)$$

for $\theta \in \Theta$.

(e) *At each juncture of $\partial \Lambda$ the normal angle jumps between the tangency angles of a convexifying bitangent.*

PROOF. Λ is convex because it is the intersection of the convex sets (halfspaces)

$$\mathcal{H}(\theta) = \{\mathbf{x} \in \mathbb{R}^2 : \mathbf{x} \cdot \mathbf{N}(\theta) \le f(\theta)\} \qquad (7.46)$$

for $\theta \in \mathbb{R}$.

By $(7.4)_4$ and (a) of (7C), the three formulas in (7.43) and (7.44) are equivalent.

We next establish (b), (c), and (e). We begin by showing that each point $\mathbf{x}$ at which $\partial \Lambda$ is smooth satisfies

$$\mathbf{x} \cdot \mathbf{N}(\theta) = f(\theta), \qquad (7.47)$$

with θ the normal angle of $\partial \Lambda$ at $\mathbf{x}$. Choose a point $\mathbf{x} \in \partial \Lambda$ at which $\partial \Lambda$ is smooth, and let $\mathbf{e} = \mathbf{x}/|\mathbf{x}|$. Then, for all sufficiently small α, $\alpha \mathbf{e} \cdot \mathbf{N}(\theta) < f(\theta)$ for all $\theta \in \mathbb{R}$. Let α_1 be the supremum of all α for which this is true, so that

$$\alpha_1 \mathbf{e} \cdot \mathbf{N}(\theta_1) = f(\theta_1) \qquad (7.48)$$

for some θ_1. Then $\alpha \mathbf{e} \cdot \mathbf{N}(\theta_1) > f(\theta_1)$ for $\alpha > \alpha_1$, so that $\alpha_1 \mathbf{e} \in \partial \Lambda$ and $\mathbf{x} = \alpha_1 \mathbf{e}$. If there is another angle θ_2 consistent with (7.48), then $\mathbf{x} \in \partial \Lambda$ must belong to the boundaries of each of the two halfspaces $\mathcal{H}(\theta_1)$ and $\mathcal{H}(\theta_2)$, so that $\mathbf{x}$ cannot be a point of smoothness of $\partial \Lambda$, a contradiction. Thus θ_1 is unique, and $\mathbf{x}$ belongs to both $\partial \Lambda$ and the boundary of a halfspace containing Λ; thus the normal of $\partial \Lambda$ must coincide with the normal $\mathbf{N}(\theta_1)$ to this halfspace. Therefore every point $\mathbf{x}$ of smoothness of $\partial \Lambda$ satisfies (7.47) with $\mathbf{N}(\theta)$ the normal to $\partial \Lambda$ at $\mathbf{x}$.

Since f is regular, it has a finite number of globally stable sections. Let Θ_i, $i = 1, 2, \ldots, N$, denote these sections, arranged in order of increasing angle; then the set Θ of globally stable angles is the union of Θ_i, and the angle jump between each pair of adjacent sections Θ_i is between the tangency angles of a convexifying bitangent.

Let $K_0(\theta)$ be defined by (7.45). Since f is regular, $f(\theta) + f''(\theta) > 0$ on Θ, so that $K_0(\theta) < 0$ is well defined on Θ. Moreover, by (1.12), $K_0(\theta)$ satisfies (1.11) when $\Theta = [0, 2\pi]$. Thus, by (1C), $K_0(\theta)$, restricted to Θ_i, is the curvature of a convex curve $\imath_i$. Let $\mathbf{r}_0(\theta)$ be defined by (7.43) for $\theta \in \Theta$. Then, by (b) of (1A), we may assume that the angle-parametrization $\mathbf{r}_i(\theta)$ of $\imath_i$ is the restriction of $\mathbf{r}_0(\theta)$ to Θ_i. Further, by $(7.4)_4$ and (c) of (7A), the function $\mathbf{r}_0(\theta)$ is continuous across the angle jumps between adjacent Θ_i; thus the countably infinite list $\imath_0 = \{\imath_1, \imath_2, \ldots\}$, with $\imath_i = \imath_{i+N}$ for all integers i, defines a PS smooth closed curve.

To complete the proof of (b), (c), and (e) it suffices to show that $\imath_0 = \partial\Lambda$ (as a curve). Choose $\theta \in \Theta$. By $(7.44)_1$, $\mathbf{r}_0(\theta) = \operatorname{grad} f(\mathbf{N}(\theta))$; hence (7.10) yields $\mathbf{r}_0(\theta) \cdot \mathbf{N}(\alpha) \leq f(\alpha)$ for all α and $\mathbf{r}_0(\theta) \in \Lambda$. But $\tau \mathbf{r}_0(\theta) \cdot \mathbf{N}(\theta) > f(\theta)$ for $\tau > 1$; hence $\mathbf{r}_0(\theta) \in \partial\Lambda$. Thus, since $\theta \in \Theta$ is arbitrary, $\imath_0 \subset \partial\Lambda$. But $\imath_0$ is a closed curve and $\partial\Lambda$ is a *simple* closed curve, and this is compatible with $\imath_0 \subset \partial\Lambda$ only if $\imath_0 = \partial\Lambda$ (as a set). Further, $\mathbf{r}_0(\theta) \cdot \mathbf{N}(\theta) = f(\theta) > 0$ for all $\theta \in \Theta$. Thus the normal $\mathbf{N}(\theta)$ to $\imath_0$ is the outward normal to $\partial\Lambda$, so that $\imath_0$ and $\partial\Lambda$ have the same orientation. Thus $\imath_0 = \partial\Lambda$ (as a curve). This completes the proof of (b), (c), and (e).

We have already established $(7.45)_1$; $(7.45)_2$ follows from (7.43) and $(1.10)_1$. Thus (a) is valid. ∎

(7Q) Equilibrium of the total energy

(a) *For $F \geq 0$ the total energy has no equilibria.*

(b) *For $F < 0$, the dilation*

$$\Omega = a\Lambda, \qquad a = |F|^{-1} \tag{7.49}$$

of the Wulff region Λ is an equilibrium; modulo translation, Ω is the only equilibrium whose boundary angle-set contains only globally stable angles.

PROOF. Assertion (a) follows from (7.34), which is necessary for an equilibrium.

To prove (b) assume that $F < 0$ and define Ω by (7.49). Then, by (b) of (7P), the angle-set of $\partial\Omega$ contains only globally stable angles, and, by (d) and (e) of (7P) and (c) of (7C), $\partial\Omega$ is consistent with the equilibrium equations (7.42), so that, by (7N), Ω is an equilibrium. We leave the remainder of the proof (uniqueness) as an exercise. ∎

Exercise Establish the assertion of uniqueness in (b) of (7Q).

7.6 Wulff's theorem

By Theorem (7L), the total energy has no minimizers within the class of bounded regions. This instability is induced by the bulk material and can be overcome by limiting the class of regions to those of given area.[28] The bulk term in (7.32) is then of no consequence and the problem reduces to

---

[28] Such variational problems arise from thermodynamical theories when the region occupied by the two phases is insulated (cf. (17.34)).

minimizing the interfacial energy

$$\mathscr{F}(\partial\Omega) = \int_{\partial\Omega} f(\theta)\,ds$$

over all bounded regions Ω of prescribed area

$$\text{area}(\Omega) = C.$$

We will refer to this problem as **Wulff's problem**. If Ω is a solution of Wulff's problem for a given value of C, then a solution for any other value of C is an appropriate dilation of Ω.

(7R) Wulff's theorem[29] *Wulff's problem has a unique*[30] *solution, and this solution is a dilation of the Wulff region (7.42).*

Wulff's theorem leads to a useful generalization of the isoperimetric ratio: given a function $e(\theta) > 0$, the **Wulff ratio** $W(e)$ **for** e is the number

$$W(e) = \frac{1}{4\pi} \frac{\left\{\int_{\partial\Lambda(e)} e(\theta)\,ds\right\}^2}{\text{area}(\Lambda(e))}, \qquad (7.50)$$

with $\Lambda(e)$ the *Wulff region* for e. By Wulff's theorem

$$\left\{\int_{\partial\Gamma} e(\theta)\,ds\right\}^2 \geq 4\pi W(e)\,\text{area}(\Gamma) \qquad (7.51)$$

for any bounded region Γ.

Exercises

1. Show that the Wulff region $\Lambda = \Lambda(f)$ satisfies:

$$\mathscr{F}(\partial\Lambda) = 2\,\text{area}(\Lambda) = 2\pi W(f).$$

2. Let $F < 0$ and define $A_0 = 4\pi W(f)/F^2$. Show that $\mathscr{E}(\Gamma) > 0$ for all bounded regions Γ with $\text{area}(\Gamma) < A_0$ (cf. Case 2 of Remark (7M)). Show further that there is a Γ with $\text{area}(\Gamma) > A_0$ and $\mathscr{E}(\Gamma) < 0$.
3. Let Ω be a solution of Wulff's problem corresponding to a given constant C. Show that Ω is an equilibrium of the total energy with $F = -[\pi W/C]^{\frac{1}{2}}$.

---

[29] A proof is beyond the scope of this book. Wulff (1901) formulated the problem and conjectured the solution (7.42). A formal proof was given by Dinghas (1944); a precise proof was given by Taylor (1978) for very general interfacial energies and a very general class of admissible regions. See also Dacorogna and Pfister (1991); Fonseca (1991); Fonseca and Müller (1991).
[30] Modulo translation.

8
EVOLUTION EQUATIONS FOR THE INTERFACE;[31] BASIC ASSUMPTIONS

To avoid repeated hypotheses, we will assume throughout the discussion of the mechanical theory that the following hypotheses are satisfied:

(8A) Assumptions

(a) *The constitutive equations are compatible with thermodynamics.*

(b) *The interfacial energy and kinetic coefficient satisfy:*

$$f(\theta) > 0, \qquad b(\theta, V) > 0,$$
$$b(\theta, V) \quad \text{is independent of } V. \tag{8.1}$$

We will refer to an evolving curve as **admissible** if it satisfies

$$b(\theta)V = [f(\theta) + f''(\theta)]K - F. \tag{8.2}$$

Evolving curves are, by definition, smooth; thus, in view of the discussion of Section 6.2, admissible curves are consistent with both balance of forces and the thermodynamically restricted constitutive equations.

8.1 Isotropic interface

For an isotropic interface f and b are constants; choosing a scaling with $f = b = 1$, (8.2) reduces to[32]

$$V = K - F. \tag{8.3}$$

A complete set of partial differential equations for an admissible evolving interface consists of (8.3) supplemented by the kinematical conditions (2.9)

---

[31] Cf. Angenent and Gurtin (1989).
[32] Mullins (1956) introduced the equation $V = K$ to study the motion of grain boundaries. Evolution according to this equation is discussed by many authors. Cf. Brakke (1978); Sethian (1985); Abresch and Langer (1986); Giga (1984, 1986); Gage and Hamilton (1986); Grayson (1987); Huisken (1987); Osher and Sethian (1988); Evans and Spruck (1991); Ilmanen (1991); and the references therein. Evans *et al.* (1991) establish the equation $V = K$ as an approximation to the Landau–Ginzburg equation, a result deduced formally by Allen and Cahn (1979) and Rubinstein *et al.* (1989).

(cf. (2.5)) satisfied by all evolving curves:

$$V = K - F, \qquad \theta_t + v\theta_s = V_s,$$
$$K_t + vK_s = V_{ss} + K^2 V, \qquad v_s = -KV \qquad (8.4)$$

(where the subscript t denotes the time derivative holding s fixed). The domains of the underlying fields in the arc-length description are not known a priori, since s varies in the interval $[0, L(t)]$ with $L(t)$ the length of the curve-set $\mathcal{A}(t)$. However, $(2.26)_2$ relates $L^{\cdot}(t)$ to KV, and we can introduce the rescaled variable $s^* = s/L(t)$.[33]

When the curve is *convex* the system (8.4) takes a particularly simple form; indeed, (8.3) and (2.20) yield

$$K_t = K^2[K_{\theta\theta} + K - F]. \qquad (8.5)$$

(with K_t the time derivative holding θ fixed).

Exercise Show that within the isotropic theory an admissible evolving curve satisfies

$$\theta^{\circ} = \theta_{ss}.$$

8.2 Anisotropic interface

8.2.1 Basic equations

For convenience, we define

$$g = b^{-1}(f + f''), \quad G = b^{-1}F; \qquad (8.6)$$

then (8.2) becomes[34]

$$V = g(\theta)K - G(\theta). \qquad (8.7)$$

A complete system of equations for an admissible evolving interface consists of (8.7) in conjunction with $(8.4)_{2-4}$. When the curve is *convex* this system reduces to

$$K_t = K^2[gK - G]_{\theta\theta} + K^2[gK - G] \qquad (8.8)$$

(with K_t the derivative holding θ fixed). This equation is also valid locally where $K \neq 0$.

---

[33] Abresch and Langer (1986).
[34] Evolution according to this equation is studied by Soner (1990); Angenent (1991); and Chen et al. (1991). The special case $V = -\psi(\theta)$ was introduced by Frank (1958).

EVOLUTION EQUATIONS FOR THE INTERFACE 61

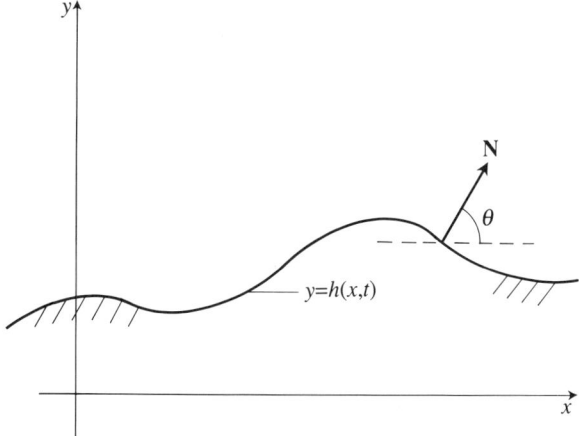

Fig. 8.1 Sign conventions when the curve is a graph $y = h(x, t)$.

(8B) Remark The term of highest order on the right side of (8.8) is $K^2 g K_{\theta\theta}$; thus (8.8) is *parabolic* for $g(\theta) > 0$, backward parabolic for $g(\theta) < 0$ (and degenerate for $K = 0$). By (7.28), (8.1), and (8.6), *parabolicity is equivalent to the strict stability of the interfacial energy, while backward parabolicity is equivalent to instability*. There is no compelling physical reason to suppose that the interfacial energy is strictly stable; in fact, material scientists often consider energies that are unstable[35] for particular ranges of the orientation θ. Since $f(\theta) > 0$ and periodic, at worst we can have an equation that is backward parabolic for some but not all values of θ.

Note that, by (8.7), the general equation (8.8), when expressed in terms of the normal velocity $V(\theta, t)$, has the form

$$g(\theta)V_t = [V + G(\theta)]^2 [V_{\theta\theta} + V]. \tag{8.9}$$

8.2.2 Equations when the interface is the graph of a function

Locally, an evolving curve may be represented as the graph of a function $y = h(x, t)$, provided the x and y axes are chosen appropriately. Consider the choice indicated in Figure 8.1 (with orientation such that arc length increases with increasing x) and let

$$w = h_x. \tag{8.10}$$

Then

$$w \tan \theta = -1, \quad h_t = (\sin \theta)^{-1} V, \quad K = h_{xx}(1 + w^2)^{-3/2}, \tag{8.11}$$

[35] Cf., for example, Gjostein (1963); Cahn and Hoffman (1974).

and the evolution equation (8.7) takes the form

$$h_t = Q(w)h_{xx} - B(w), \tag{8.12}$$

$$Q(w) = g(\theta)\sin^2\theta, \quad B(w) = \frac{G(\theta)}{\sin\theta}, \quad \tan\theta = -w^{-1},$$

or, differentiating with respect to x,

$$w_t = [Q(w)w_x - B(w)]_x, \tag{8.13}$$

which is in conservation form.

8.2.3 Equations when the interface is a level set

The zero-level set of a function $\varphi(\mathbf{x}, t)$ defines a family of evolving sets

$$\mathscr{A}(t) = \{\mathbf{x} \in \mathbb{R}^2 : \varphi(\mathbf{x}, t) = 0\},$$

and if grad $\varphi(\mathbf{x}, t) \neq 0$ on $\mathscr{A}(t)$, we may consider $\mathscr{A}(t)$ as an evolving curve. Choosing a parametrization such that the normal $\mathbf{N}(\mathbf{x}, t)$ to $\mathscr{A}(t)$ points in the direction of increasing φ, then

$$\mathbf{N} = \frac{\operatorname{grad}\varphi}{|\operatorname{grad}\varphi|} \tag{8.14}$$

and the corresponding normal speed and curvature are given by

$$\begin{aligned}V &= -\varphi_t|\operatorname{grad}\varphi|^{-1},\\ K &= -|\operatorname{grad}\varphi|^{-3}\{\varphi_{xx}(\varphi_y)^2 - 2\varphi_{xy}\varphi_x\varphi_y + \varphi_{yy}(\varphi_x)^2\}.\end{aligned} \tag{8.15}$$

Suppose that $\mathscr{A}(t)$ is governed by the evolution equation (8.7), and consider $g(\theta)$ and $G(\theta)$ as functions $g(\operatorname{grad}\varphi)$ and $G(\operatorname{grad}\varphi)$ (cf. (8.14), (1.3)). Then, by (8.15), we may replace (8.7) by the partial differential equation[36]

$$\varphi_t = g(\operatorname{grad}\varphi)|\operatorname{grad}\varphi|^{-2}\{\varphi_{xx}(\varphi_y)^2 - 2\varphi_{xy}\varphi_x\varphi_y + \varphi_{yy}(\varphi_x)^2\}$$
$$+ G(\operatorname{grad}\varphi)|\operatorname{grad}\varphi|, \tag{8.16}$$

which is well defined not only on the evolving curve $\varphi = 0$, but also away from this curve; in fact, at all points and times for which grad $\varphi \neq 0$.

The problem of finding an admissible evolving curve with prescribed initial value $\mathscr{A}(0)$ is formally equivalent to finding a solution φ of (8.16) with $\mathscr{A}(0) = \{\mathbf{x} : \varphi(\mathbf{x}, 0) = 0\}$. This formulation in terms of level sets is important because:

(1) it removes the necessity of tracking the curve;

---

[36] This idea is due to Sethian (1985), Barles (1985), and Osher and Sethian (1988), who restrict attention to the isotropic equation (8.2). Soner (1990) and Chen et al. (1991) consider the anisotropic equation.

(2) it allows for a larger class of 'evolving curves';

(3) it allows for a weak formulation[37] in the sense of viscosity solutions.[38]

Exercise Establish the relations (8.14)–(8.16).

8.3 Plan of the next few chapters

In the next two chapters we will restrict attention to interfacial energies f that are strictly stable; for such energies the interface is necessarily smooth (cf. (7.30)). In Chapter 11 we will allow f to be unstable, but we will utilize corners in the interface to remove the unstable angle intervals. In Chapter 12 we will allow f to be both unstable and non-smooth.

---

[37] Cf. Soner (1990); Chen *et al.* (1991); Evans and Spruck (1991).
[38] Crandall and Lions (1983).

9

STATIONARY INTERFACES AND STEADILY EVOLVING INTERFACES[39]

We assume throughout this chapter that *the interfacial energy is strictly stable*, so that

$$f(\theta) + f''(\theta) > 0$$

for all θ. We write

$$W(\theta) = [f(\theta) + f''(\theta)]^{-1}; \tag{9.1}$$

then, by (1.12),

$$\int_0^{2\pi} W(\theta)^{-1} e^{i\theta} \, d\theta = 0. \tag{9.2}$$

This chapter is restricted to evolving interfaces that are admissible; to avoid repetition, *we will omit the term 'admissible' in most of the ensuing discussion*.

Evolving interfaces are, by definition, *smooth*, a limitation that involves no loss in generality, since the interfacial energy is strictly stable (cf. (7.30)).

9.1 Stationary interfaces

By a **stationary interface** we mean an evolving interface that is independent of time. A trivial consequence of (8.7) is that for $F = 0$ the unbounded time-independent facets form the complete collection of stationary interfaces.

For $V = 0$, (8.2) coincides with the equilibrium equation $(7.36)_1$. Further, since the interface is necessarily smooth, we need not worry about $(7.36)_2$. Thus, by (7N), *the boundary $\partial\Omega$ of a bounded region Ω (which without loss in generality we take to be occupied by phase 1) is a stationary interface if and only if Ω is an equilibrium of the total energy* (7.32). We may therefore conclude from (7N) that there are no such stationary interfaces if $F \geq 0$. On the other hand, by (7N) and the strict stability of the interfacial energy, there is a unique bounded stationary interface $\partial\Omega$ for any $F < 0$; in fact, $\Omega = |F|^{-1} \Lambda$ with Λ the Wulff region (7.42), and the angle-parametrization of $\partial\Omega$ is $|F|^{-1}$ times (7.43) (or (7.44)).

---

[39] Cf. Angenent and Gurtin (1989).

STATIONARY AND STEADILY EVOLVING INTERFACES   65

9.2 Steadily evolving facets

By (2.30), (8.2), and (9.1), steadily evolving interfaces satisfy

$$K(\theta) = [F + b(\theta)\mathbf{U} \cdot \mathbf{N}(\theta)]W(\theta), \tag{9.3}$$

where $\mathbf{U}$ ($\neq \mathbf{0}$) is the steady velocity. It is convenient to introduce the vector function

$$\mathbf{b}(\theta) = b(\theta)\mathbf{N}(\theta), \tag{9.4}$$

whose locus forms the *polar diagram* Polar(b) of b, and to write (9.3) in the form

$$K(\theta) = \Phi(\theta)W(\theta), \qquad \Phi(\theta) = F + \mathbf{U} \cdot \mathbf{b}(\theta). \tag{9.5}$$

Note that a steadily evolving interface is a steadily evolving facet if and only if the corresponding normal angle θ is identically constant and a solution of

$$\Phi(\theta) = 0. \tag{9.6}$$

The equation (9.6) has a simple geometric solution. To state this solution concisely, we introduce the following terminology. For $\mathbf{d} \neq \mathbf{0}$, let $\ell(\mathbf{d})$ denote the straight line

$$\ell(\mathbf{d}) = \{\mathbf{x}: \mathbf{d} \cdot \mathbf{x} = |\mathbf{d}|^2\}; \tag{9.7}$$

equation (9.7) defines a one-to-one correspondence between non-zero vectors and lines that do not pass through the origin; we will refer to $\mathbf{d}$ as the **support vector** for $\ell = \ell(\mathbf{d})$. In the same spirit, for $\mathbf{d} \neq \mathbf{0}$, we write $\ell = \ell(0\mathbf{d})$ for the line through the origin perpendicular to $\mathbf{d}$ and refer to ℓ as the line with **support vector** $0\mathbf{d}$. Then

$$\begin{array}{c}\Phi(\theta_0) = 0 \text{ if and only if the line with} \\ \text{support vector } (-F/|\mathbf{U}|^2)\mathbf{U} \text{ intersects} \\ \text{the polar diagram Polar}(b) \text{ at } \mathbf{b}(\theta_0),\end{array} \tag{9.8}$$

and we have the following result (Figure 9.1).

(9A) Theorem *Given any vector* $\mathbf{U} \neq \mathbf{0}$, *there is a steadily evolving facet with steady velocity* $\mathbf{U}$ *and normal angle* θ_0 *if and only if the line with support vector* $(-F/|\mathbf{U}|^2)\mathbf{U}$ *intersects* Polar(b) *at* θ_0.

9.3 Steadily evolving interfaces that are not flat

Let F and $\mathbf{U} \neq \mathbf{0}$ be given. Let $\mathscr{s}$ be a steadily evolving interface corresponding to F with $\mathbf{U}$ as steady velocity, and assume that $\mathscr{s}$ is **not flat** in the sense

66 STATIONARY AND STEADILY EVOLVING INTERFACES

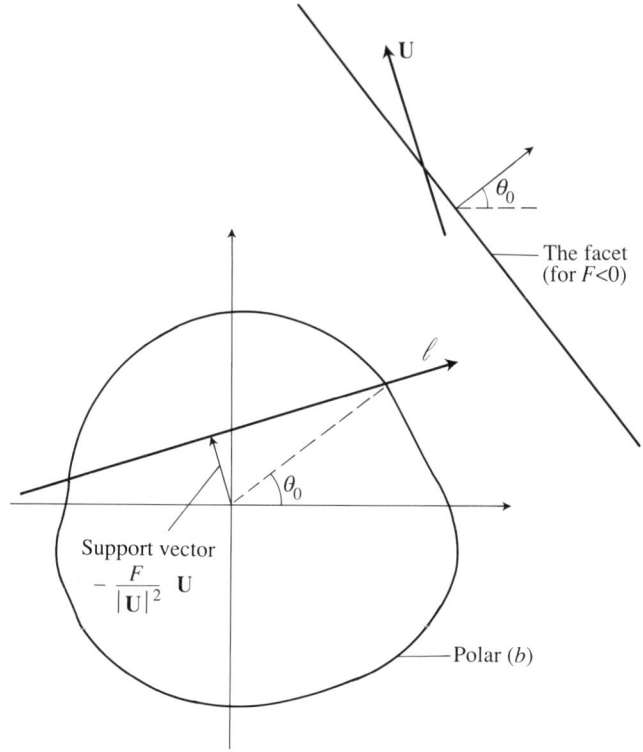

Fig. 9.1 When the line ℓ with support vector $(-F/|\mathbf{U}|^2)\mathbf{U}$ intersects Polar(b) at θ_0, then there is a steadily evolving facet with steady velocity $\mathbf{U}$ and direction θ_0.

that its curvature is not identically zero. Then (9.3) and Lemma 1B imply that $\jmath$ is *convex*. By definition, $\jmath(t)$ is a boundary curve at each t, and hence is simple and either closed or unbounded. Assume that $\jmath$ is closed. Let $U = |\mathbf{U}|$, $\mathbf{e} = \mathbf{U}/U$. Then (1.11) and (9.5) yield

$$\int_0^{2\pi} [\Phi(\theta)W(\theta)]^{-1}\mathbf{e}\cdot\mathbf{N}(\theta)\,d\theta = 0. \tag{9.9}$$

Let $M(\theta, \mathbf{e}, U)$ denote the left side of (9.9). By (9.2) and $(9.5)_2$, $M(\theta, \mathbf{e}, 0) = 0$. Further, differentiating $M(\theta, \mathbf{e}, U)$ with respect to U yields the conclusion that $M(\theta, \mathbf{e}, U)$ is strictly monotone in U. Thus (9.9) is possible only if $U = 0$, which violates the definition of a steadily evolving interface; hence $\jmath$ cannot be closed. Thus $\jmath$ is *unbounded*.

Therefore, appealing to Lemma 1C, the angle-set of $\jmath$ is a bounded interval (θ_1, θ_2), and this interval and the accompanying curvature $K(\theta)$ must be a

STATIONARY AND STEADILY EVOLVING INTERFACES   67

solution of the problem:

> Find an angle interval (θ_1, θ_2), $\theta_2 - \theta_1 \leq \pi$,
> such that $K(\theta)$, defined by (9.3), is non-vanishing   (9.10)
> on (θ_1, θ_2) and has $K(\theta_1) = K(\theta_2) = 0$.

Conversely, if (θ_1, θ_2) is consistent with (9.10), then Lemma 1C of Section 1.2 implies that $K(\theta)$ restricted to (θ_1, θ_2) is the curvature of a convex bounded curve $\mathcal{J}_0$. Let $\mathcal{J}$ denote the steadily evolving interface whose portrait is $\mathcal{J}_0$ and whose steady velocity is **U** (cf. (2.28)); trivially, $\mathcal{J}$ has $K(\theta)$ as its curvature.

Thus we are reduced to solving (9.10); since $W(\theta) > 0$, (θ_1, θ_2) is a solution of (9.10) if and only if

$$\theta_1, \theta_2, \quad 0 < \theta_1 - \theta_2 \leq \pi, \quad \text{are } \textit{consecutive} \text{ zeros of } \Phi(\theta). \quad (9.11)$$

To facilitate the discussion of such zeros, let us agree to call a line ℓ a **chord for** Polar(b) **between** θ_1 **and** θ_2 if $0 < \theta_1 - \theta_2 \leq \pi$ and ℓ intersects Polar(b) at $\mathbf{b}(\theta_1)$ and $\mathbf{b}(\theta_2)$, but not at any other point $\mathbf{b}(\theta)$ with $\theta \in (\theta_1, \theta_2)$. In view of (9.8), (9.11) is then equivalent to the requirement that:

> The line ℓ with support vector $(-F/|\mathbf{U}|^2)\mathbf{U}$
> be a chord for Polar(b) between θ_1 and θ_2.

Given a line ℓ with ℓ a chord for Polar(b) between θ_1 and θ_2, we write

$$\ell|_{(\theta_1, \theta_2)} := \{\mathbf{x}: \mathbf{x} = \mathbf{b}(\theta_1) + \alpha[\mathbf{b}(\theta_2) - \mathbf{b}(\theta_1)], \alpha \in (0,1)\}$$

for the segment of ℓ between $\mathbf{b}(\theta_1)$ and $\mathbf{b}(\theta_2)$.

Continuing as before, let F and $\mathbf{U} \neq \mathbf{0}$ be given, and let $\mathbf{r}$ be a steadily evolving interface corresponding to F and $\mathbf{U}$. Let (θ_1, θ_2) be the angle-set for $\mathcal{J}$, so that the line ℓ with support vector $(-F/|\mathbf{U}|^2)\mathbf{U}$ is a chord for Polar(b) between θ_1 and θ_2. Choose $\theta \in (\theta_1, \theta_2)$. Then there is a unique point $\mathbf{x} \in \ell|_{(\theta_1,\theta_2)}$ such that $\mathbf{x} = \alpha \mathbf{N}(\theta)$ for some $\alpha \geq 0$; in addition,

$$b(\theta) > \alpha \quad \text{or} \quad b(\theta) < \alpha \quad \text{according as } \ell|_{(\theta_1, \theta_2)} \quad (9.12)$$
is interior or exterior to Polar(b).

Since $\mathbf{x} \in \ell$, $\mathbf{x} \cdot \mathbf{U} = -F$; thus (9.4) and (9.5) yield

$$\Phi(\theta) = [b(\theta) - \alpha]\mathbf{U} \cdot \mathbf{N}(\theta).$$

Further $K = W\Phi$ with $W > 0$ and $K(\theta) \neq 0$; hence

$$K(\theta)\mathbf{U} \cdot \mathbf{N}(\theta) = C(\theta)[b(\theta) - \alpha], \quad C(\theta) > 0,$$

and we conclude, with the aid of (2.31), (2.32), and (9.12), that $\mathcal{J}$ is a *steadily evolving bump*, which recedes or advances according as $\ell|_{(\theta_1,\theta_2)}$ is interior or exterior to Polar(b).

The results established above are summarized in the next theorem, in which F and $\mathbf{U} \neq \mathbf{0}$ are assumed prescribed.

(9B) Theorem on steadily evolving interfaces Let ℓ denote the line with support vector $(-F/|\mathbf{U}|^2)\mathbf{U}$. Let $\mathscr{s}$ be a non-flat steadily evolving interface that corresponds to F and has $\mathbf{U}$ as steady velocity. Then $\mathscr{s}$ is a steadily evolving bump. If (θ_1, θ_2) is the angle-set of $\mathscr{s}$, then ℓ is a chord for Polar(b) between θ_1 and θ_2, and $\mathscr{s}$ is receding or advancing according as $\ell|_{(\theta_1, \theta_2)}$ is interior or exterior to Polar(b).

Conversely, if ℓ is a chord for Polar(b) between θ_1 and θ_2, then there is a unique steadily evolving interface $\mathscr{s}$ corresponding to F with (θ_1, θ_2) as angle-set and $\mathbf{U}$ as steady velocity.

(9C) Corollary *There are no advancing bumps if* Polar(b) *is convex, and none when* $F = 0$.

(9D) Remarks
1. For an isotropic material ($f = b = 1$) with $F = 0$, (9.3) reduces to $K(\theta) = \mathbf{U} \cdot \mathbf{N}(\theta)$; letting $\mathbf{U} = U(1, 0)$, $U > 0$, there are two steadily evolving interfaces, one with angle-set $(-\pi/2, \pi/2)$, the other with angle-set $(\pi/2, 3\pi/2)$, and both are receding bumps.[40]

2. It is generally believed that dendritic growth requires diffusion in the bulk material. It is interesting that a steadily advancing bump is possible even without diffusion. In the present theory such growth is a consequence of anisotropy in the kinetic coefficient and results when certain orientations suffer drag forces sufficiently lower than neighbouring orientations.

---

[40] This solution, due to Mullins (1956), is referred to as the 'grim reaper' by geometers (cf. Grayson (1987), p. 298).

10
GLOBAL BEHAVIOUR FOR AN INTERFACE WITH STABLE ENERGY[41]

In this chapter we analyse the global behaviour of the interface under the assumption of a *strictly stable* interfacial energy, and therefore consider evolving interfaces that are *admissible* in the sense of the general anisotropic equation

$$b(\theta)V = [f(\theta) + f''(\theta)]K - F \qquad (10.1)$$

with

$$f(\theta) + f''(\theta) > 0, \qquad b(\theta) > 0 \qquad (10.2)$$

for all $\theta \in \mathbb{R}$ (cf. (8.1), (7.28)).

10.1 Existence of evolving interfaces from a prescribed initial curve

The evolution problem of interest consists of tracing an interface evolving from a given initial configuration. More precisely, we have the following problem.

(10A) Evolution problem Given f, b, and an *initial region* Ω_0, find $\Omega(t)$ for $t \geq 0$ such that $\Omega(0) = \Omega_0$ and $\Omega(t)$ is the reference region of an admissible evolving interface.

(10B) Existence theorem *Let f and b be C^∞ (and consistent with (10.2)). Let Ω_0 be a given initial domain, which we assume to be bounded with boundary a Lipschitz-continuous simple closed curve. Then there is a unique, maximal family of domains $\Omega(t)$ ($0 \leq t < T_{\max}$) such that:*

(a) $\partial \Omega(t)$, *for $t > 0$, is a C^2 simple, closed curve, continuous in t for $0 \leq t < T_{\max}$;*

(b) *the evolution of $\partial \Omega(t)$ is governed by (10.1);*

(c) $\Omega(0) = \Omega_0$.

In fact, this solution is C^∞ for $0 < t < T_{\max}$.

---
[41] Cf. Angenent and Gurtin (1989).

This theorem is due to Angenent (1991),[42] who shows that, for $T_{max} < \infty$, one of the following must be true.

(1) $\sup_{s \in \mathbb{R}} |K(s, t)| \to \infty$ as $t \to T_{max}$.

(2) K and its derivatives remain bounded as $t \to T_{max}$, so that $\partial\Omega(t)$ converges to a C^∞ curve Γ; however, Γ is not simple.

Condition (1) will occur whenever the interface shrinks to zero, or whenever the interface develops a kink; condition (2) indicates the formation of self-intersections or self-tangencies.

When Ω_0 is smooth, $\partial\Omega(t)$ admits a parametrization $(p, t) \mapsto \mathbf{r}(p, t)$ as an admissible evolving interface of duration T_{max}. In the next section we will study the behaviour of such interfaces as measured by their **perimeter** $L(t)$ and **enclosed area** $A(t)$,

$$L(t) = \text{length}(\partial\Omega(t)), \qquad A(t) = \text{area}(\Omega(t)). \tag{10.3}$$

We will however, restrict our attention to evolving interfaces, termed regularly maximal, whose singularity at $t = T_{max}$ (for $T_{max} < \infty$) is not too pathological.

Precisely, an admissible evolving interface with duration T is **regularly maximal** if either $T = \infty$ or

$$T < \infty \quad \text{and} \quad A(t) \to 0 \quad \text{or} \quad L(t) \to \infty \quad \text{as } t \to T. \tag{10.4}$$

Regularly maximal evolving interfaces cannot be extended beyond $t = T$, but for T finite they exhibit fairly regular behaviour as $t \to T$: they either explode or disappear. This class of evolving interfaces does not include interfaces that develop self-tangencies, self-intersections, or kinks at $t = T$.

10.2 Growth and decay of the interface

Let $\mathscr{J}$ denote a *bounded, admissible* evolving interface. Then $\mathscr{J}(t)$ encloses a bounded region $\Omega(t)$; without loss of generality, we assume that the normal $\mathbf{N}$ to $\mathscr{J}$ is the outward normal to $\partial\Omega(t)$. Then:

$$\text{As } s \text{ increases from } 0 \text{ to } L(t),$$
$$\theta \text{ decreases from } 0 \text{ to } -2\pi. \tag{10.5}$$

We write

$$\mathscr{F}(t) = \int_{\partial\Omega(t)} f(\theta)\, ds \tag{10.6}$$

---

[42] Cf. Soner (1990); Chen et al. (1991).

for the **total interfacial energy**. The next result, essentially the dissipation inequality, is an immediate consequence of (5.4) and the discussion of the paragraph containing (6.9).

(10C) Growth theorem[43]

$$\mathcal{F}^{\bullet}(t) + F A^{\bullet}(t) = -\int_{\partial \Omega(t)} b(\theta) V^2 \, ds \leq 0. \tag{10.7}$$

Exercise Prove (10.7) directly by differentiating $\mathcal{F}(t)$ and using (10.1).

The inequality $F > 0$ occurs when $\Omega(t)$ has higher bulk energy than its exterior; in this instance (10.7) indicates a tendency for the less stable reference phase to shrink. On the other hand, $F < 0$ when the reference phase has lower bulk energy; here $FA(t)$ is negative and of the wrong sign for a Lyapunov function, indicating a tendency for the more stable reference phase to grow, at least in situations for which area dominates length. The next theorem shows that this is indeed the case. In fact, we show that for $F \geq 0$ the reference phase shrinks to zero; for $F < 0$ the reference phase shrinks to zero when initially small, but grows unboundedly when initially large. These conclusions are in accord with the discussion of Remark (7M).

(10D) Theorem on the growth of the refernce phase *Consider an evolving interface that has duration T and is regularly maximal and admissible.*

(a) *If $F \geq 0$, then $T < \infty$ and $A(t) \to 0$ as $t \to T$.*

(b) *If $F < 0$, then:*

    (i) *if $L(0)$ is sufficiently small, then $T < \infty$ and $A(t) \to 0$ as $t \to T$;*

    (ii) *if $A(0)$ is sufficiently large, then $T = \infty$ and $A(t) \to \infty$ as $t \to \infty$.*

PROOF. For any 2π-periodic function $\varphi(\theta)$ on $\mathbb{R}$, let

$$\varphi_{\text{av}} = \frac{1}{2\pi} \int_0^{2\pi} \varphi(\theta) \, d\theta, \quad \varphi_{\text{max}} = \sup_{\theta \in \mathbb{R}} \varphi(\theta), \quad \varphi_{\text{min}} = \inf_{\theta \in \mathbb{R}} \varphi(\theta). \tag{10.8}$$

If we use (1.5) and (10.5) to change variable in $(10.8)_1$ from θ to s, we arrive at

$$\int_{\partial \Omega(t)} \varphi(\theta) K \, ds = -2\pi \varphi_{\text{av}}. \tag{10.9}$$

Note that, by (10.2), $g_{\text{av}} > 0$, $g_{\text{min}} > 0$.

---

[43] Gurtin (1988).

It is convenient to rewrite (10.1) in the form

$$V = g(\theta)K - G(\theta), \tag{10.10}$$

$$g(\theta) = \frac{f(\theta) + f''(\theta)}{b(\theta)}, \qquad G(\theta) = \frac{F}{b(\theta)}.$$

We begin with the identities:

$$A^{\cdot}(t) = -2\pi g_{\mathrm{av}} - F \int_{\partial\Omega(t)} b(\theta)^{-1} \, ds,$$

$$L^{\cdot}(t) = -2\pi G_{\mathrm{av}} - \int_{\partial\Omega(t)} g(\theta)K^2 \, ds. \tag{10.11}$$

To derive $(10.11)_1$ integrate $(10.10)_1$ over $\partial\Omega(t)$ and use $(2.26)_1$ and (10.9); to derive $(10.11)_2$ multiply $(10.10)_1$ by K, integrate over $\partial\Omega(t)$, and use $(2.26)_2$ and (10.9).

If we can show that $L(t) \to 0$ in finite time provided the solution persists that long, then we can conclude from (10.4) that $T < \infty$ and $A(t) \to 0$ as $t \to T$; we cannot conclude (from this alone) that $L(t) \to 0$ as $t \to T$.

Assume that $F \geq 0$. Then, by $(10.11)_2$ and the remark made in the previous paragraph, (a) follows.

Assume that $F < 0$. By (10.6), $f_{\min} L \leq \mathscr{F} \leq f_{\max} L$; thus (10.7) and $(10.11)_1$ yield

$$\mathscr{F}^{\cdot} \leq -FA^{\cdot} \leq -2\pi|F|g_{\mathrm{av}} + \left(\frac{F^2}{f_{\min} b_{\min}}\right)\mathscr{F},$$

which implies (b)(i).

Next, $(10.11)_2$ yields

$$L(t) \leq L(0) + 2\pi|F|(b^{-1})_{\mathrm{av}} t. \tag{10.12}$$

On the other hand, (7.51) yields

$$\int_{\partial\Omega(t)} b(\theta)^{-1} \, ds \geq 2[\pi W(b^{-1})A(t)]^{\frac{1}{2}},$$

with $W(b^{-1})$ the Wulff region for b^{-1}; therefore, by $(10.11)_1$,

$$A^{\cdot}(t) \geq DA(t)^{\frac{1}{2}} - C,$$

$$C = 2\pi g_{\mathrm{av}} > 0, \qquad D = 2|F|[\pi W(b^{-1})]^{\frac{1}{2}} > 0. \tag{10.13}$$

By (10.4), (10.12), and (10.13), if $A(0)$ is sufficiently large, then $T = \infty$ and $A(t) \to \infty$ as $t \to \infty$. ∎

(10E) Remarks Assume that $F < 0$. The following remarks follow from the proof of the last theorem.

1. If
$$L(0) < \frac{2\pi g_{av} b_{min} f_{min}}{|F| f_{max}},$$

then $A(t) \to 0$; if
$$A(0) > \frac{\pi (g_{av})^2}{F^2 W(b^{-1})}, \tag{10.14}$$

then $A(t) \to \infty$. Thus for an *isotropic material*: the interface will shrink provided the initial length of the interface is less than that of a circle of radius $f/|F|$; the interface will grow without bound provided the reference region has initial area larger than that of a circle of radius $f/|F|$.

2. For *the interface initially a circle*,
$$A(t) \to 0 \quad \text{if } L(0) < \delta_0 \ell, \qquad A(t) \to \infty \quad \text{if } L(0) > \ell,$$
$$\ell = \frac{2\pi b_{max} g_{av}}{|F|}, \qquad \delta_0 = \frac{b_{min} f_{min}}{b_{max} f_{av}}. \tag{10.15}$$

ℓ represents a *critical circumference* for a circular interface, while $\delta_0 \in (0, 1]$ is a measure of the underlying *anisotropy*; for initial circumferences between $\delta_0 \ell$ and ℓ, (10.15) furnishes no information. For an *isotropic material*, $\delta_0 = 1$, and the reference phase grows or shrinks according as the the initial circumference is greater than or less than $\ell = 2\pi f/|F|$.

10.3 Evolution of curvature; fingers

The next theorem shows that the total curvature between inflection points cannot increase.

(10F) Theorem *Let $\imath$ be an admissible evolving curve. Assume that the curvature does not change sign on $\imath$ and vanishes at the end points of $\imath$. Then*[44]
$$\frac{d}{dt} \int_{\imath(t)} |K| \, ds \leq 0. \tag{10.16}$$

In fact, the angle-set $\Theta(t)$ of $\imath(t)$ nests as t increases.[45]

---

[44] Cf. Brakke (1978), Prop. 2, p. 230 and Abresch and Langer (1986) for the case $V = K$.
[45] Cf. Grayson (1987), Lemma 1.9(iii), for the case $V = K$. The term '$\Theta(t)$ nests' means $\Theta(t_2) \subset \theta(t_1)$ whenever $t_2 > t_1$.

PROOF. Let $[S_1(t), S_2(t)]$ denote the arc-length interval corresponding to $\jmath(t)$. We will give a proof for $K(s, t) \geq 0$ on $[S_1(t), S_2(t)]$. (The proof for $K \leq 0$ is analogous.) For any function $\varphi(s, t)$, let $\varphi_i(t) = \varphi(S_i(t), t)$. Then, by hypothesis,

$$K_i(t) = 0. \tag{10.17}$$

Further, since $K = \theta_s$,

$$\frac{d}{dt} \int_{\jmath(t)} |K| \, ds = \theta_2^{\bullet}(t) - \theta_1^{\bullet}(t), \tag{10.18}$$

and, by (2.16) and (10.17), $\theta_i^{\bullet} = (V_s)_i$. To complete the proof, it therefore suffices to show that

$$(V_s)_1 \geq 0, \qquad (V_s)_2 \leq 0. \tag{10.19}$$

But $V = g(\theta)K - G(\theta)$, so that $V_s = g(\theta)K_s + g'(\theta)K^2 - G'(\theta)K$, and $(K_s)_1 \geq 0$, $(K_s)_2 \leq 0$, since $K \geq 0$ with $K = 0$ at the endpoints; thus (10.19) follows. ∎

By (10.9), a convex evolving interface $\jmath$ satisfies

$$\frac{d}{dt} \int_{\jmath(t)} |K| \, ds = 0.$$

A counterpart of this result for a non-convex interface is a consequence of the next theorem.

(10G) Theorem[46] *Let $\jmath$ denote a bounded, admissible evolving interface. Then*

$$\frac{d}{dt} \int_{\jmath(t)} |K| \, ds \leq 0. \tag{10.20}$$

Further, the number of inflection points[47] *of $\jmath(t)$ cannot increase with t.*

This theorem follows from results of Angenent and Gurtin [1992]. A formal justification of (10.20) follows upon dividing the interface into subcurves on which K does not change sign, and then appealing to (10.16) on each subcurve.

Theorem (10G) has the following important corollary.

(10H) Corollary *An initially convex admissible evolving interface will remain convex for all time.*

---

[46] Cf. Abresh and Langer (1986) for the case $V = K$.
[47] We use the term **inflection point** for a point at which the curvature changes sign; this definition makes sense: from the parabolicity of (8.12), straight line segments disappear immediately.

(10I) Remark Roughly speaking, a **finger** may be defined as a section of the interface between inflection points. Theorem (10G) and (10.16) then have the following corollary:

> The total number of fingers as well as the total curvature of each finger cannot increase with time. (10.21)

The next result, essentially a consequence of (8.12) and the comparison principle[48] for parabolic equations, shows that nested interfaces remain nested.

(10J) Containment principle[49] *Let $\Omega(t)$ and $\underline{\Omega}(t)$, $0 \leq t < T$, be bounded reference regions for admissible evolving interfaces (corresponding to the same b, f, and F). Assume that the closure of $\Omega(0)$ is contained in the interior of $\underline{\Omega}(0)$. Then the closure of $\Omega(t)$ is contained in the interior of $\underline{\Omega}(t)$ for $0 \leq t < T$.*

For the *isotropic* equation $bV = fK - F$ (b and f constant) an interface initially a circle will not change its shape as it evolves. Wulff regions are a natural analogue—within the *anisotropic* theory—of the circle and, as the next result shows, when the kinetic coefficient $b(\theta)$ is proportional to $f(\theta)^{-1}$, Wulff regions evolve without change in shape. In this result and the next $\Lambda(f)$ is the Wulff region for the energy f (cf. (7.42)); we will also utilize the Wulff region $\Lambda(b^{-1})$, with b the kinetic coefficient.

(10K) Soner's solution[50] *Consider the special case in which*

$$b(\theta)f(\theta) = \kappa \ (=\text{constant}) \tag{10.22}$$

for all $\theta \in \mathbb{R}$. Let

$$\Omega(t) = z(t)\Lambda(f),$$

$z(t) > 0$, $0 \leq t < T$, so that $\Omega(t)$ is a dilation of the Wulff region $\Lambda(f)$. Then $\Omega(t)$ is the reference region of an admissible evolving interface if and only if $z(t)$ is a solution of the differential equation

$$\kappa z^{\cdot}(t) = -z(t)^{-1} - F. \tag{10.23}$$

PROOF. Since $\Omega(t)$ is convex, we may use (θ, t) as independent variables in describing $\partial \Omega(t)$. The angle-parametrizations $\mathbf{r}(\theta, t)$ and $\mathbf{r}_0(\theta)$ of $\partial \Omega(t)$ and $\partial \Lambda$ are related by

$$r(\theta, t) = z(t)\mathbf{r}_0(\theta),$$

---
[48] Cf. Protter and Weinberger (1967).
[49] Cf. Angenent and Gurtin (1989, 1992); Soner (1990); Chen *et al.* (1991).
[50] Soner (1990).

and therefore, by (1.10) (for $\mathbf{r}_0$) and (7.43),

$$V(\theta, t) = \mathbf{N}(\theta) \cdot \mathbf{r}_t(\theta, t) = z^{\cdot}(t)f(\theta).$$

Similarly, the curvatures $K(\theta, t)$ and $K_0(\theta)$ of $\partial\Omega(t)$ and $\partial\Lambda$ are related by $K(\theta, t) = z(t)^{-1}K_0(\theta)$, and therefore, by (7.45),

$$K(\theta, t) = -z(t)^{-1}[f(\theta) + f''(\theta)]^{-1}.$$

Thus, by (10.22), the evolution equation (10.1) is equivalent to (10.23). ∎

(10L) Remarks Solutions $z(t)$ of (10.23) have the following asymptotic properties (cf. (10D)):

1. If $F \geq 0$ or if $F < 0$ and $z(0) < |F|^{-1}$, then $z(t)$ decreases monotonically to zero in finite time.

2. If $F < 0$ and $z(0) > |F|^{-1}$, then $z(t)$ increases monotonically to infinity as $t \to \infty$; in fact,

$$z(t) = \kappa^{-1}|F|t + O(1) \qquad \text{as } t \to \infty. \tag{10.24}$$

In this case, by (10.22), $\Lambda(f) = \kappa\Lambda(b^{-1})$ and we may use (10.24) to conclude that $\Omega(t)$ is asymptotic to $t|F|\Lambda(b^{-1})$ as $t \to \infty$ (that is, the Hausdorff distance between $t^{-1}\Omega(t)$ and $|F|^{-1}\Lambda(b^{-1})$ tends to zero as $t \to \infty$).

Theorem (10D) asserts that for $F < 0$ and the interface of sufficiently large area, the interface grows unboundedly. The next result gives the corresponding asymptotic behaviour of the interface.

(10M) Asymptotic behaviour[51] Let $F < 0$. Let $\Omega(t)$, $0 \leq t < \infty$, be the reference region for an admissible evolving interface with

$$|F|^{-1}\Lambda(f) \subset \Omega(0). \tag{10.25}$$

Then $\text{area}(\Omega(t)) \to \infty$ as $t \to \infty$; in fact, $\Omega(t)$ is asymptotic to $t|F|\Lambda(b^{-1})$.

(10N) Remarks

1. The estimate (10.14) is in some cases better than (10.25). For example, granted isotropy (10.25) implies unbounded growth if $\Omega(0)$ contains a circle of radius $f/|F|$, but (10.14) yields growth as long as $A(0)$ is larger than the area of this circle, irrespective of the shape of $\Omega(0)$.

---

[51] A slightly weaker version of this result was conjectured by Angenent and Gurtin (1989) and proved by Soner (1990). Soner's proof uses his exact solution (10K) and a stronger version of the containment principle involving subsolutions and supersolutions. The result as stated here is given by Angenent and Gurtin (1992).

2. The asymptotic behaviour of the interface for those cases in which it shrinks to zero (parts (a) and (b)(i) of (10D)) remains an open question.

3. The asymptotic shape $t|F|\Lambda(b^{-1})$ will have corners if the polar diagram of $b(\theta)$ is not convex (cf. (b) of (7P) with $\Lambda = \Lambda(b^{-1})$).

11

UNSTABLE INTERFACIAL ENERGIES AND INTERFACES WITH CORNERS[52]

Suppose now that

the interfacial energy is regular, but not stable

(cf. the last paragraph of Section 7.3). Then an admissible evolving interface must, by virtue of its *smoothness*, involve normal angles at which the evolution equations are backward parabolic. Two ways of overcoming this difficulty are to:

(1) regularize the evolution equations;

(2) allow corners that correspond to jumps in θ across the unstable portions of $f(\theta)$.

In this chapter we will consider the second alternative. In particular, we will discuss piecewise-smooth (PS) evolving curves whose angle-sets consist of globally stable angles,[53] a condition that leads to *parabolic* evolution equations. What makes such curves possible is the use of corners to eliminate normal angles that are not globally stable. In Section 11.5 we will extend the theory to all normal angles by allowing the interface to wrinkle infinitesimally on angle-sets that are not globally stable.

11.1 Admissibility; corner conditions

Our interest is in PS evolving curves that are consistent with balance of forces and correspond to normal angles that are *globally stable*. Precisely, a PS evolving curve $\mathscr{a}$ is **admissible**[54] if:

(A1) $\mathscr{a}$ is genuine (cf. (2.35));

(A2) at each time, the angle-set of $\mathscr{a}$ contains only globally stable angles;

---

[52] Cf. Angenent and Gurtin (1989).

[53] For the growth of a solid seed in its melt, one might assume that the seed is initially stable in the sense that it minimizes total interfacial energy at fixed area. Granted this, the seed will be a dilation of the Wulff region for f (cf. (7R)), so that initially, and *presumably locally in time*, the angle-set of the boundary will be limited to globally stable angles (cf. (b) of (7P)).

[54] In previous sections, which concerned *smooth* evolving curves, admissibility meant consistency with balance of forces (and the underlying constitutive equations). Here admissibility signifies, in addition, the requirement that the underlying angle-set contain only *globally stable angles*.

(A3) the capillary force $\mathbf{C}(\theta)$ defined by (6.10) is consistent with *balance of forces*

$$\int_{\partial\imath(t)} \mathbf{C}(\theta) = \int_{\imath(t)} (F + b(\theta)V)\mathbf{N}\,ds \qquad (11.1)$$

(cf. (6.11)) whenever $\imath$ is an evolving subcurve of $\jmath$.

Balance of forces requires that a corner defined by a jump in normal angle from θ^- to θ^+ satisfy $\mathbf{C}(\theta^-) = \mathbf{C}(\theta^+)$ (cf. Section 6.2). With this in mind, we define an **admissible corner** to be a pair $\{\theta^-, \theta^+\}$ of *distinct globally stable angles* consistent with $\mathbf{C}(\theta^-) = \mathbf{C}(\theta^+)$. By (c) of (7C) and the regularity of f,

$\{\theta^-, \theta^+\}$ is an admissible corner if and only
if θ^- and θ^+ are tangency angles of a
convexifying bitangent to the Frank diagram. \qquad (11.2)

(11A) Local force–balance relations Let $\jmath$ be a PS evolving curve consistent with (A2). Then $\jmath$ is admissible if and only if:

(A4) *each of its arcs $\jmath_i$ evolves according to*

$$V = g(\theta)K - G(\theta),$$
$$g(\theta) = \frac{f(\theta) + f''(\theta)}{b(\theta)}, \qquad G(\theta) = \frac{F}{b(\theta)}; \qquad (11.3)$$

(A5) *at each juncture i, the normal angles θ_i^- and θ_i^+ are independent of time with $\{\theta_i^-, \theta_i^+\}$ an admissible corner.*

PROOF. Since $\jmath$ satisfies (A2), we have only to show that (A1), (A3) $\Leftrightarrow$ (A4), (A5). In view of the discussion given in the paragraph containing (6.13), the conditions (A4) and

$$\mathbf{C}(\theta_i^-(t)) = \mathbf{C}(\theta_i^+(t)) \qquad \text{at each juncture } i \qquad (11.4)$$

are together equivalent to (A3). Further, (A5) $\Rightarrow$ (A1). Thus (A4), (A5) $\Rightarrow$ (A1), (A3). Conversely, suppose that (A1), (A3) hold, so that (A4) and (11.4) are satisfied. Let i be an arbitrary juncture. By (A1), there is a time t_0 such that $\theta_i^-(t_0) \neq \theta_i^+(t_0)$, so that, by (6.13), (c) of (7C), and the regularity of f, $\mathbf{C}(\theta_i^-(t_0)) = \mathbf{C}(\theta_i^+(t_0))$ and θ_i^- and θ_i^+ are tangency angles of a convexifying bitangent to the Frank diagram. We may therefore use (7.31) and the smoothness of $\theta_i^-(t)$ and $\theta_i^+(t)$ (which follows from the smoothness of the individual arcs) to conclude that $\theta_i^-(t) \equiv \theta_i^-(t_0)$ and $\theta_i^+(t) \equiv \theta_i^+(t_0)$. ∎

A direct consequence of (A2) and (A5) is the following.

(11B) Proposition *The angle-set of an admissible PS evolving interface is the complete set of globally stable angles.*

Admissible PS evolving curves that are *convex* are best described using (θ, t) as independent variables, since the evolution equations then have simple forms, and since the values of θ at each juncture are independent of time. For such curves, the evolution of each arc must be consistent with the equation (cf. (8.9))

$$g(\theta)V_t = [V + G(\theta)]^2[V_{\theta\theta} + V], \tag{11.5}$$

while the relevant jump conditions follow from the requirement that (cf. (2.40))

$$V\mathbf{N} - V_\theta \mathbf{T} \quad \text{be continuous across each juncture.} \tag{11.6}$$

By $(2.36)_1$ and (A5), the juncture curvatures ϑ_i of an admissible PS evolving curve cannot vanish; thus, for such a curve to be convex, each of its evolving arcs must be convex and, at each juncture i, the curvatures K_i^+, K_i^-, and ϑ_i must have the same sign.

(11C) Theorem *Let $\mathscr{a}$ be an admissible PS evolving curve. Then given any juncture i of $\mathscr{a}$, K_i^+, K_i^-, and ϑ_i are all ≥ 0 or all ≤ 0. Thus $\mathscr{a}$ is convex if and only if each of its arcs is convex.*

PROOF. Let S denote the arc length of a juncture i of $\mathscr{a}$ at a given time t. The corner at i corresponds to a convexifying bitangent to the Frank diagram; let (α, β), $0 < \beta - \alpha < \pi$, denote the angle interval defined by this bitangent. The two possibilities for the corner at i are $\theta_i^- = \alpha$, $\theta_i^+ = \beta$ and $\theta_i^- = \beta$, $\theta_i^+ = \alpha$. Consider the former. Since $\theta(s)$ lies outside of (α, β), $\theta(s) \leq \alpha$ for $s \leq S$ and $\theta(s) \geq \beta$ for $s \geq S$, so that $K_i^\pm(t) \geq 0$. Further, by (2.36), the juncture curvature at i is $\vartheta_i = \beta - \alpha > 0$. Thus K_i^+, K_i^-, and ϑ_i are all ≥ 0. The proof for $\theta_i^- = \beta$, $\theta_i^+ = \alpha$ is left as an exercise. ∎

Exercise Complete the proof.

11.2 The initial-value problem

The evolution problem of interest—stated in Section 10.1 for stable energies—here has the following form:

(11D) Evolution problem Given f, b, and an *initial region* Ω_0, find $\Omega(t)$ for $t \geq 0$ such that $\Omega(0) = \Omega_0$ and $\Omega(t)$ is the reference region of an admissible PS evolving interface.

To ensure that this problem is meaningful, we assume that:

(P) the angle-set of $\partial\Omega_0$ contains only globally stable angles, and the points of $\partial\Omega_0$ at which the normal angle jumps correspond to admissible corners.

Then the angle-set of $\partial\Omega_0$ (and hence the angle-set of the subsequent evolution) is

Θ, the set of globally stable angles,

and each fixed angle pair that defines a juncture (corner) for the subsequent evolution is an angle jump between adjacent connected components of Θ. Moreover, for Ω_0 convex we can use (11.5) and (11.6) to rephrase the problem as an initial-value problem for the velocity $V(\theta, t)$, at least as long as $\Omega(t)$ remains convex. The relevant initial data $V_0(\theta)$ is determined using (11.3):

$$V_0(\theta) = g(\theta)K_0(\theta) - G(\theta) \tag{11.7}$$

for $\theta \in \Theta$, where $K_0(\theta)$ is the curvature of $\partial\Omega_0$. We are therefore led to the following problem.

(11E) Evolution problem for convex interfaces Given f, b, and the initial velocity $V_0(\theta_0)$ for $\theta_0 \in \Theta$, find a solution $V(\theta, t)$ of the system

$$g(\theta)V_t = [V + G(\theta)]^2[V_{\theta\theta} + V] \quad \text{for } \theta \in \Theta, t > 0,$$

$$V\mathbf{N} - V_\theta \mathbf{T} \quad \text{continuous across each juncture for } t > 0, \tag{11.8}$$

$$V(\theta, 0) = V_0(\theta) \quad \text{for } \theta \in \Theta.$$

Here the initial data V_0 should be compatible (in the sense of (11.7)) with a convex initial region Ω_0 consistent with (P).

The convex evolution problem is well posed locally in time, the only difficulty being a degeneracy in the equation (11.8)$_1$ at normal angles θ for which $[V + G(\theta)] = 0$, a condition that signifies $K = 0$. An interesting question appropriate to this problem concerns whether or not convexity is preserved for all time.

Exercise Establish the analogue of Soner's solution (10K) for an interfacial energy that is not stable.

11.3 Facets and wrinklings that connect evolving curves

Let $\mathfrak{d}$, $\mathfrak{d}_A$, and $\mathfrak{d}_B$ be admissible PS evolving curves. Then $\mathfrak{d}$ **admissibly connects** $\mathfrak{d}_A$ **and** $\mathfrak{d}_B$ provided $\{\mathfrak{d}_A, \mathfrak{d}, \mathfrak{d}_B\}$ is also an admissible PS evolving curve. Tacit in this definition is the requirement (cf. (A1)) that the composite curve be non-smooth at the **connection junctures** (that is, at the junctures between ∂_A and ∂ and between ∂_B and ∂); in fact, by (A5), these junctures necessarily correspond to admissible corners.

(11F) Facet decay theorem *Let $L(t)$ be the length of an admissible facet that admissibly connects admissible PS evolving curves. Then $L'(t) \leq 0$. If the curvature of ∂_A or ∂_B at its juncture with ∂ is non-zero, then $L'(t) < 0$.*

The facet decay theorem follows from (11C) and the next proposition.

(11G) Proposition *Let ∂ be an admissible facet that admissibly connects admissible PS evolving curves ∂_A and ∂_B. For $C = A, B$, let θ_C and K_C denote the normal angle and curvature of ∂_C at its juncture with ∂, and let ϑ_C designate the corresponding juncture curvature. Then*

$$\theta_A = \theta_B, \qquad \vartheta_A = -\vartheta_B, \tag{11.9}$$

with θ_A and the normal angle of the facet both tangency angles for a convexifying bitangent to the Frank diagram. Further, the length $L(t)$ of the facet satisfies

$$L' = \frac{g(\theta_A)[K_A - K_B]}{\sin \vartheta_A}. \tag{11.10}$$

PROOF. On an admissible facet, $\theta(t) = \theta(s, t)$ is independent of s, so that, by (11.3), $V = -G(\theta(t))$ and $V_s = 0$. Thus, by (2.9)$_2$, $\theta° = 0$; since $\theta_t = \theta° - v\theta_s$, $\theta_t = 0$. Thus

$$\theta \equiv \text{constant on an admissible facet.} \tag{11.11}$$

The connection junctures correspond to the corners $\{\theta_A, \theta\}$ and $\{\theta, \theta_B\}$, and, by (A5) (Section 11.1), θ_A, θ, and θ_B must be tangency angles of a convexifying bitangent; thus by (2.36)$_1$, (11.9) holds.

If we apply (2.41)$_1$ to ∂_A and ∂, and (2.42)$_1$ to ∂ and ∂_B, and subtract the two relations, we find, using (2.18) and (11.9), that $V_A - V_B = -L' \sin \vartheta_A$; in view of (11.3), (11.10) follows. ∎

(11H) Remark The assumed *regularity* of f is necessary for the validity of both this proposition and the facet decay theorem: if f is not regular, we can conclude that θ_A, θ, and θ_B lie on the same multitangent, but this is not enough to prove either (11.9) or (11.10).

(11I) Properties of admissible wrinklings *Any wrinkling whose normal angles are tangency angles of a convexifying bitangent to the Frank diagram is* *admissible; these are the only admissible, fully faceted evolving curves.*[55] *Let ∂ be such a wrinkling, with α and β the corresponding normal angles. Then ∂ has the following properties*:

(a) *The normal velocity is given by*

$$V = -b(\alpha)^{-1}F \quad \text{on the facets with normal angle } \alpha,$$
$$V = -b(\beta)^{-1}F \quad \text{on the facets with normal angle } \beta.$$

(b) $\mathbf{C}(\theta)$ *has the constant value* $\mathbf{C}(\alpha) = \mathbf{C}(\beta)$ *on ∂*.

(c) *The juncture velocity* **w** *of any juncture is a constant independent of the particular juncture in question, so that the system of internal facets behaves as a rigid body translating with constant velocity* **w**. *Further*,

$$\mathbf{w}\cdot\mathbf{N}(\alpha) = -b(\alpha)^{-1}F, \quad \mathbf{w}\cdot\mathbf{N}(\beta) = -b(\beta)^{-1}F. \qquad (11.12)$$

PROOF. Let ∂ be a wrinkling whose normal angles α and β are tangency angles of a convexifying bitangent to the Frank diagram. Then (a) and (b) are satisfied, and this implies (A4) and (A5), so that, by (11A), ∂ is admissible. Further, (c) follows from (a) and the linear independence of $\mathbf{N}(\alpha)$ and $\mathbf{N}(\beta)$. To complete the proof we have only to show that such wrinklings are the only admissible, fully faceted evolving curves; but this follows from (11.11) and (11G). ∎

(11J) Remark Consider the evolution problem with initial data consistent with property (P). Suppose that $\partial\Omega_0$ contains a convex section Γ_0 that is connected to facets at its initial and terminal points. Let Y_0 denote the angle-set of Γ_0. Since Γ_0 is convex, the *internal junctures* of Γ_0 correspond to the jumps in θ between adjacent connected components of Y_0. Let θ_A and θ_B, respectively, denote the normal angles corresponding to the initial and terminal points of Γ_0, let $\{\gamma_A, \theta_A\}$ and $\{\theta_B, \gamma_B\}$ denote the admissible corners at these endpoints, and let $\vartheta_A = \gamma_A - \theta_A$ and $\vartheta_B = \theta_B - \gamma_B$ designate the corresponding juncture curvatures. The subsequent evolution of Γ_0 can then be described in terms of the velocity $V(\theta, t)$ using analogues of (11.8) together with 'boundary conditions' at the initial and terminal junctures found using (2.12), (2.21), (2.41)$_1$, and (2.42)$_1$. In fact, the resulting system is given by:

$$g(\theta)V_t = [V + G(\theta)]^2[V_{\theta\theta} + V] \quad \text{for } \theta \in Y_0, \, t > 0,$$

$$V\mathbf{N} - V_\theta\mathbf{T} \quad \text{continuous across each internal juncture for } t > 0,$$

$$V(\theta_A, t)\cos\vartheta_A + V_\theta(\theta_A, t)\sin\vartheta_A = -G(\gamma_A),$$

$$V(\theta_B, t)\cos\vartheta_B - V_\theta(\theta_B, t)\sin\vartheta_B = -G(\gamma_B),$$

$$V(\theta, 0) = V_0(\theta) \quad \text{for } \theta \in Y_0$$

---

[55] The smoothness of $f(\theta)$ is crucial. In Chapter 12 we will show that certain non-smooth energies exhibit more general types of facetings.

This system yields a (locally) well-posed problem for $V(\theta, t)$. What is interesting is that *the facets decouple the evolution of $V(\theta, t)$ (and hence $K(\theta, t)$) for $\theta \in Y_0$ from the evolution of the remainder of the interface*. If the initial interface $\partial\Omega_0$ consists of convex sections separated by wrinkles and facets, then the above procedure gives $V(\theta, t)$ and $K(\theta, t)$ for each convex section; with this informaton, the evolution of the facets and wrinklings (from prescribed initial positions) is easily determined using (11G) and (c) of (11I). Granted this is done, the initial and terminal positions of the convex sections are known as functions of time, and these data and a knowledge of the corresponding curvatures yield the complete evolutionary behaviour of the convex sections, and hence of the complete interface.

11.4 Equations near a corner when the curve is a graph

Consider an admissible PS evolving curve represented in a neighbourhood of an admissible corner $\{\theta^-, \theta^+\}$ as the graph of a function $y = h(x, t)$. Here $x = \zeta(t)$ is the position at time t of the corner, and the curve is oriented so that arc length increases with increasing x (cf. Figure 8.1). Then $h(x, t)$ is continuous and piecewise smooth, with a jump discontinuity in $w = h_x$ at the **free boundary** $x = \zeta(t)$.

The function w satisfies (8.13) away from the free boundary and is consistent with two free-boundary conditions. The first of these, a direct consequence of (8.11), is given by

$$w(\zeta(t) \pm 0, t) = P^\pm \qquad (11.13)$$

with

$$P^\pm = -\cot \theta^\pm.$$

The second condition is more complicated. For any function $\varphi(x, t)$, let $\varphi^\pm(t) = \varphi(\zeta(t) \pm 0, t)$. By $(2.39)_1$,

$$V^\pm = \mathbf{R}^{\cdot} \cdot \mathbf{N}^\pm; \qquad (11.14)$$

thus, since $\mathbf{N} = (\cos\theta, \sin\theta)$, if we eliminate the y-component of $\mathbf{R}^{\cdot}$ between the two equations (11.14), and use the fact that $\zeta^{\cdot}$ is the x-component of $\mathbf{R}^{\cdot}$, we arrive at an expression for $\zeta^{\cdot}$ as a function of $V^\pm$ and $\theta^\pm$; the expressions $V = gK - G$ and $K = w_x(1 + w^2)^{-3/2}$ then lead to the free-boundary condition:

$$\zeta^{\cdot} = A^+(w_x)^+ - A^-(w_x)^- - C \qquad (11.15)$$

with $A^\pm$ and C the constants defined by

$$A^\pm = \frac{g(\theta^\pm)D^\pm}{W^\pm}, \qquad C = G(\theta^+)A^+ - G(\theta^-)A^-.$$

$$W^\pm = (1 + (P^\pm)^2)^{3/2}, \qquad D^+ = \frac{\sin \theta^-}{a}, \qquad D^- = \frac{\sin \theta^+}{a},$$

$$a = \sin(\theta^- - \theta^+).$$

The basic system of equations then consists of (8.13) away from $x = \zeta(t)$ supplemented by (11.13) and (11.15) at $x = \zeta(t)$. A change in dependent variable renders this system more transparent. Let

$$u(x, t) = \begin{cases} A^-(w(x, t) - P^-) & \text{for } x < \zeta(t) \\ A^+(w(x, t) - P^+) & \text{for } x > \zeta(t), \end{cases}$$

so that $u(x, t)$ is continuous across $x = \zeta(t)$. Further, let Q and B be as specified in (8.12), and define

$$Q^\pm(u) = Q((u/A^\pm) + P^\pm).$$

Then the system under consideration reduces to the partial differential equations

$$\begin{aligned} u_t &= [Q^-(u)u_x - B^-(u)]_x & \text{for } x < \zeta(t), \\ u_t &= [Q^+(u)u_x - B^+(u)]_x & \text{for } x > \zeta(t), \end{aligned} \qquad (11.16)$$

and the free-boundary conditions

$$\begin{aligned} u(\zeta(t) \pm 0, t) &= 0, \\ u_x(\zeta(t) + 0, t) - u_x(\zeta(t) - 0, t) &= \zeta'(t) + C. \end{aligned} \qquad (11.17)$$

Apart from the constant C, which can be transferred from (11.17) to (11.16) by the coordinate change $x^* = x + Ct$, (11.17) are *exactly the free-boundary conditions of the classical Stefan problem*.

11.5 Interfaces with arbitrary angle-set; infinitesimal wrinklings

Thus far we have restricted attention to interfaces with globally stable angle-sets. We now present a method of discussing interfaces whose angle-sets contain angles that are not globally stable, perhaps even unstable.

Through this section:

(α, β), $0 < \beta - \alpha < \pi$, is the angle interval of
a convexifying bitangent to the Frank diagram, (11.18)

so that $\{\alpha, \beta\}$ and $\{\beta, \alpha\}$ are admissible corners. Consider a portion $\mathscr{J}$ of an evolving interface such that $\mathscr{J}$ has normal angles in the interval (α, β), but the remainder of the interface does not, at least near $\mathscr{J}$. Then, by (7K) and the regularity of f, the angle-set of $\mathscr{J}$ might contain unstable angles. A method of treating $\mathscr{J}$ without having to consider unstable normal angles is to assume that ∂ *immediately forms an infinitesimally wrinkled curve*[56] with normal angles α and β (cf. Section 1.5); since its facets are *infinitesimally small*, the wrinkled curve coincides with the interface. A justification of this assumption is that wrinkling the curve, even infinitesimally, lowers the energy (cf. Remark (7I)). Since the properties (a)–(c) of (11I) of admissible wrinklings are independent of the particular wrinkling (with normal angles α and β), we adopt them as defining conditions.

Precisely, an **admissible infinitesimal wrinkling** with normal angles α and β is a PS evolving curve ω such that:

(W1) the angle-set of ω lies in the interval (α, β);

(W2) ω evolves as a rigid body translating with constant velocity **w**, and **w** is consistent with

$$\mathbf{w} \cdot \mathbf{N}(\alpha) = -b(\alpha)^{-1} F, \qquad \mathbf{w} \cdot \mathbf{N}(\beta) = -b(\beta)^{-1} F; \qquad (11.19)$$

however, we do not require that the *endpoints* of ω evolve with velocity **w**: ω is allowed to shrink or grow tangentially.

The **corresponding capillary force** is then *defined* to have the constant value

$$\mathbf{C}(\theta) = \mathbf{C}(\alpha) = \mathbf{C}(\beta) \qquad \text{on } \omega, \qquad (11.20)$$

and the kinematic parameters such as **V**, **N**, and **T** that describe the motion are taken to be those of the actual evolving curve ω.

The evolution of wrinklings is best described using the **effective kinetic modulus**[57] $b_{\text{eff}}(\theta)$, which is the continuous (2π-periodic) function on $\mathbb{R}$ defined as follows (Figure 11.1): $b_{\text{eff}}(\theta) = b(\theta)$ for all globally stable angles θ; the polar diagram of $b_{\text{eff}}(\theta)$ is a straight line between adjacent globally stable sections. The interval (α, β) is a typical angle interval between globally stable sections, and on such intervals $b_{\text{eff}}(\theta)$ has the explicit form

$$b_{\text{eff}}(\theta)^{-1} = \mu_\alpha(\theta) b(\alpha)^{-1} + \mu_\beta(\theta)(\beta)^{-1} \qquad (11.21)$$

for $\alpha \leq \theta \leq \beta$, where $\mu_\alpha(\theta)$ and $\mu_\beta(\theta)$ are the facet densities defined by (cf. (1.19))

$$\mathbf{T}(\theta) = \mu_\alpha(\theta) \mathbf{T}(\alpha) + \mu_\beta(\theta) \mathbf{T}(\beta).$$

To verify (11.21), consider $\varphi(\theta) = b_{\text{eff}}(\theta)^{-1}$ as an 'interfacial energy', so that Polar(b_{eff}) is the corresponding 'Frank diagram'. Since this Frank diagram is a straight line between α and β, (11.21) follows from (c) of (7D).

The next result justifies the term 'effective kinetic modulus'.

---

[56] This idea is due to Cahn and Taylor (private communication).
[57] Polar(b_{eff}) is convex if Polar(b) is convex. It is important to note that—in contrast to $f^{\#}$—the 'Frank diagram' of b_{eff} is generally *not* convex. It is important to note that b_{eff} depends not only on b, but also on f.

UNSTABLE INTERFACIAL ENERGIES

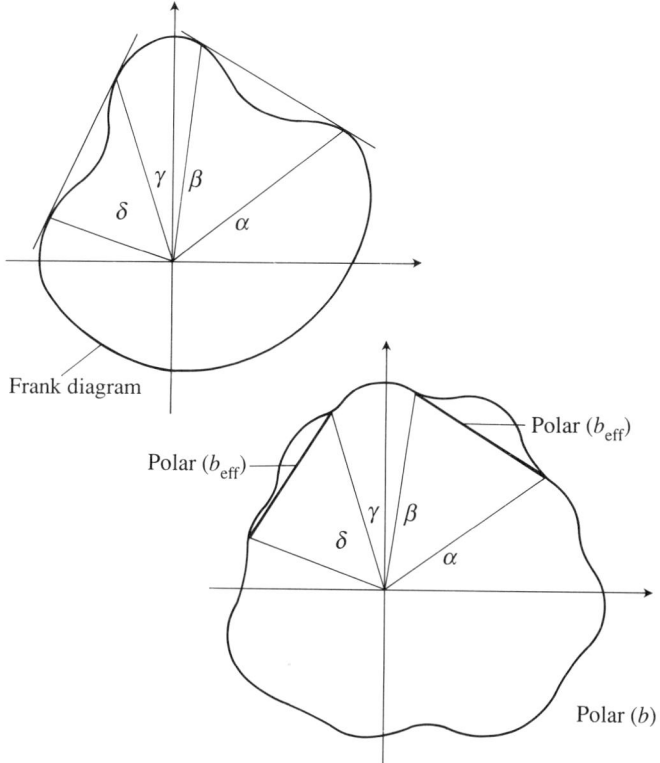

Fig. 11.1 Construction of the effective kinetic modulus $b_{\text{eff}}(\theta)$. Given a Frank diagram with (α, β) and (γ, δ) each an angle interval between adjacent globally stable sections, Polar (b_{eff}) are constructed by replacing Polar (b) between α and β and between γ and δ by straight lines, as shown.

(11K) Theorem *Admissible infinitesimal wrinklings evolve according to*

$$b_{\text{eff}}(\theta)V = -F. \qquad (11.22)$$

PROOF. Consider an admissible infinitesimal wrinkling with normal angles α and β. If $F = 0$, then, by (11.19), $\mathbf{w} = 0$, so that $V = \mathbf{w} \cdot \mathbf{N} = 0$. Thus (11.22) is satisfied. Assume that F is non-zero. Then, since $\mathbf{N}(\theta) = \mu_\alpha(\theta)\mathbf{N}(\alpha) + \mu_\beta(\theta)\mathbf{N}(\beta)$, we may use (11.19) and (11.21) to conclude that

$$V = \mathbf{w} \cdot \mathbf{N}(\theta) = -F[\mu_\alpha(\theta)b(\alpha)^{-1} + \mu_\beta(\theta)b(\beta)^{-1}] = -Fb_{\text{eff}}(\theta). \quad \blacksquare$$

Let ω be an admissible infinitesimal wrinkling, and let $\mathfrak{I}_C$ and $\mathfrak{I}_D$ be admissible PS evolving curves. We say that ω **admissibly connects** $\mathfrak{I}_C$ and $\mathfrak{I}_D$ provided $\{\mathfrak{I}_C, \omega, \mathfrak{I}_D\}$ is a PS evolving curve with capillary force *continuous* across the two connection junctures. Let α and β denote the normal angles of ω. Then, since the capillary force $\mathbf{C}(\theta)$ can equal $\mathbf{C}(\alpha)$ or $\mathbf{C}(\beta)$ only at

$\theta = \alpha$ or $\theta = \beta$ (cf. (11.2)), we may use (11.20) to conclude that:

(W3) the normal angle of the terminal point of ∂_C is α or β, and similarly for the initial point of ∂_D.

In addition, by (W1),[58] the normal angle $\theta(s, t)$ of ω can never take on the values α or β, and therefore *the composite curve necessarily suffers a jump in normal angle at each of the connection junctures.*

(11L) Decay theorem for infinitesimal wrinklings *Let ω be an admissible infinitesimal wrinkling that admissibly connects admissible PS evolving curves ∂_C and ∂_D. Then*

$$\omega(t_2) \subset \omega(t_1) + (t_2 - t_1)\mathbf{w} \qquad \text{for } t_2 > t_1, \tag{11.23}$$

so that, modulo a rigid translation, $\omega(t)$ nests as t increases. Further, the inclusion in (11.23) is strict if the curvature of ∂_C or ∂_D at its juncture with ω is non-zero.

PROOF. Consider the juncture between ∂_C and ω. Let θ_C, $V_C(t)$, $v_C(t)$, $K_C(t)$, $\mathbf{T}_C$, $\mathbf{N}_C$, respectively, denote the normal angle, normal velocity, tangential endpoint velocity, curvature, tangent, and normal for ∂_C at this juncture, let $\theta_w(t)$, $V_w(t)$, $v_w(t)$, $\mathbf{T}_w(t)$, $\mathbf{N}_w(t)$ denote corresponding quantities for ω, and let ω denote the rigid velocity of ω. Then $v_w(t) - \mathbf{w}\cdot\mathbf{T}_w(t)$ represents the tangential velocity of the juncture relative to the tangential component of the rigid velocity of the wrinkling. To establish the result (11.23) it suffices to show that

$$v_w(t) \geq \mathbf{w}\cdot\mathbf{T}_w(t), \tag{11.24}$$

that the inequality is strict if the curvature of ∂_C at the juncture with ω is non-zero, and that the relative tangential velocity of the juncture between ω and ∂_D satisfies analogous inequalities with $\geq$ and $>$ replaced by $\leq$ and $<$. We will establish only those results concerning (11.24).

By $(2.41)_1$ and $(2.42)_2$,

$$v_w = V_C \sin \vartheta + v_C \cos \vartheta, \qquad v_C = -V_w \sin \vartheta + v_w \cos \vartheta,$$

where $\vartheta \neq 0$ is the juncture curvature of the juncture in question. Thus

$$v_w = (\sin \vartheta)^{-1}\{V_C - V_w \cos \vartheta\}$$

---

[58] This assumption is made for technical reasons. If the normal angle is continuous across the connection juncture C with ∂_C, say, and the curvature of ∂_C at C is non-zero, then the normal velocity is discontinuous across C, indicating a tendency of the normal angle to develop a discontinuity. To verify that this actually happens requires a weak formulation of the problem.

and, since $V_w = \mathbf{w} \cdot \mathbf{N}_w$,

$$v_w - \mathbf{w} \cdot \mathbf{T}_w = (\sin \vartheta)^{-1}\{V_C - \mathbf{w} \cdot \mathbf{N}_w \cos \vartheta - \mathbf{w} \cdot \mathbf{T}_w \sin \vartheta\}.$$

But $\mathbf{N}_C = \mathbf{T}_w \sin \vartheta + \mathbf{N}_w \cos \vartheta$. Thus

$$\begin{aligned} v_w - \mathbf{w} \cdot \mathbf{T}_w &= (\sin \vartheta)^{-1}\{V_C - \mathbf{w} \cdot \mathbf{N}_C\} \\ &= (\sin \vartheta)^{-1}\{V_C - \mathbf{w} \cdot \mathbf{N}_C\} \\ &= (\sin \vartheta)^{-1}\{g(\theta_C)K_C - G(\theta_C) + G(\theta_C)\} \\ &= (\sin \vartheta)^{-1}g(\theta_C)K_C. \end{aligned} \quad (11.25)$$

Let α and β (consistent with (11.18)) denote the normal angles of ω. By (W3), the angle θ_C is either α or β; assume $\theta_C = \alpha$. Then, by (W1), (11.18), and $(2.36)_1$, $\vartheta > 0$. Further, since the angle set of $\mathfrak{d}_C$ contains only globally stable angles, we may conclude from (11.18) that $\theta(s, t) \leq \alpha$ on $\mathfrak{d}_C$, so that $K_C \geq 0$. Thus, since $g(\theta_C) > 0$, (11.25) is ≥ 0, with the inequality strict if $K_C > 0$. If $\theta_C = \beta$, then $\vartheta < 0$, $K_C \leq 0$, and we have the same conclusion regarding (11.25). ∎

Exercise Supply the missing steps in the proof above by analysing the juncture of $\mathfrak{d}_D$ with ω.

A **generalized, admissible PS evolving interface** is a PS evolving interface $\mathfrak{d}$ that is the union of

admissible PS evolving curves $\imath_p$, $p = 1, \ldots, P$,

and

admissible infinitesimal wrinklings ω_p, $p = 1, \ldots, P$,

such that the $\imath_p$'s alternate with the ω_p's, and such that each ω_p admissibly connects the PS evolving curves adjoining it.

The evolution of such generalized curves admits a simple description using the *effective kinetic modulus* $b_{\text{eff}}(\theta)$ in conjunction with the *convexified energy* $f^\#(\theta)$. Indeed, since $f^\#(\theta)$ coincides with $f(\theta)$ whenever θ is globally stable, but has $f^{\#\prime\prime}(\theta) + f^\#(\theta) = 0$ otherwise (as the convexified Frank diagram is flat between the tangency angles of each convexifying bitangent (cf. (7D)), we are led to the following consequence of (11.3) and (11.22).

(11M) Theorem *Generalized, admissible PS evolving interfaces evolve according to*

$$b_{\text{eff}}(\theta)V = [f^{\#\prime\prime}(\theta) + f^\#(\theta)]K - F. \quad (11.26)$$

(11N) Remark The evolution equation (11.26) has the form

$$V = g_{\text{eff}}(\theta)K - G_{\text{eff}}(\theta), \quad (11.27)$$

with $g_{\text{eff}}(\theta) = [f^{\#''}(\theta) + f^{\#}(\theta)]/b_{\text{eff}}(\theta)$, $G_{\text{eff}}(\theta) = F/b_{\text{eff}}(\theta)$. This equation is *parabolic* at normal angles that are globally stable, but is otherwise *hyperbolic*, with $g_{\text{eff}}(\theta)$ generally discontinuous at the corresponding transition.[59]

11.6 Stationary interfaces and steadily evolving interfaces with corners

Admissible PS stationary interfaces and admissible PS steadily evolving interfaces are defined as for smooth interfaces (Chapter 9), except that we now add the requirements that the angle-set contain only globally stable angles and that all corners be admissible. To avoid repetition, we will omit the term 'admissible' in the ensuing discussion.

The discussion of Section 9.1 is valid here almost without change. In particular, the boundary $\partial\Omega$ of a bounded region Ω (occupied by phase 1) is a stationary interface if and only if Ω is an equilibrium of the total energy (7.32). For $F \geq 0$ there are no stationary interfaces. For $F < 0$ there is a unique bounded stationary interface $\partial\Omega$; $\Omega = |F|^{-1}\Lambda$ with Λ the Wulff region (7.42); the angle-parametrization of $\partial\Omega$ is $|F|^{-1}$ times (7.43).

We also have the possibility of steadily evolving interfaces with corners. Let $F \neq 0$ be given. Let $\{\theta_1, \theta_2\}$ be an admissible corner with θ_1 and θ_2 globally stable, let ℓ be a line with ℓ a chord for Polar(b) between θ_1 and θ_2 (if b is strictly convex, then exactly one such line exists), and compute **U** by the requirement that $-(F/|\mathbf{U}|^2)\mathbf{U}$ be the support vector for ℓ (cf. Section 9.2). Then, any (stationary) infinite wrinkling in which θ jumps back and forth between θ_1 and θ_2 is a portrait of a steadily evolving interface with steady velocity **U**. One might refer to this evolution as a *steady wrinkling*.

Other solutions are possible. Angenent and Gurtin (1989) construct steadily receding and steadily advancing bumps with corners for Frank diagrams and b-diagrams of certain prescribed shapes.

---

[59] The equation (11.27) is studied by Gurtin *et al.* (1992).

12
NON-SMOOTH INTERFACIAL ENERGIES; CRYSTALLINE ENERGIES[60]

Material scientists often consider interfacial energies that are continuous but have derivatives that suffer jump discontinuities.[61] We now discuss a particular class of energies with this property. These energies, termed crystalline,[62] have convexified Frank diagrams that are polygonal. As we shall see, crystalline energies are compatible with fully faceted interfaces.

12.1 Crystalline energies

We assume that the energy $f(\theta)$ is **crystalline**; precisely, we assume that $f(\theta)$ is continuous, piecewise-smooth, strictly positive, and consistent with the following two conditions.

(E1) The convexification $C(\mathscr{G})$ of the Frank diagram $\mathscr{G}$ is polygonal.

(E2) The vertices of this polygon are the only points at which $\mathscr{G}$ touches $C(\mathscr{G})$.

The angles that define the vertices will be referred to as the **preferred orientations** of $f(\theta)$, and preferred orientations that define adjacent vertices will be termed **adjacent**. We write

Θ for the set of preferred orientations.

By (E1) and (E2), the preferred orientations are the only globally convex angles of the Frank diagram; unfortunately, *they also represent points of discontinuity for $f'(\theta)$*.

Consider the convexified Frank diagram $C(\mathscr{G})$ as the Frank diagram of the convexified energy $f^{\#}(\theta)$. Then $C(\mathscr{G})$ is the locus of the vector function

$$\gamma(\theta) = f^{\#}(\theta)^{-1}\mathbf{N}(\theta), \qquad (12.1)$$

which we will refer to as the **convexified Frank potential**. By (E2),

$$f(\theta) = f^{\#}(\theta) \quad \text{for } \theta \in \Theta, \qquad f(\theta) > f^{\#}(\theta) \quad \text{for } \theta \notin \Theta. \qquad (12.2)$$

---

[60] Cf. Angenent and Gurtin (1989).
[61] Cf., e.g., Herring (1951a,b); Cahn and Hoffman (1974).
[62] Cf. Wulff (1901). See also Taylor (1978).

12.2 The Wulff region

The Wulff region Λ, given by (7.42), is well defined for the crystalline energy $f(\theta)$; as for a smooth energy, the unique solution of Wulff's problem for $f(\theta)$ is a suitable dilation of Λ.

(12A) Properties of the Wulff region

(a) Λ *is a convex polygon.*

(b) *The angle-set of $\partial \Lambda$ is the set Θ of preferred orientations.*

(c) *For each $\theta \in \Theta$, the facet of $\partial \Lambda$ with normal angle θ is a segment of the line $\{\mathbf{x}: \mathbf{x} \cdot \mathbf{N}(\theta) = f(\theta)\}$.*

PROOF. We first prove that

$$\Lambda = \bigcap_{\theta \in \Theta} \mathcal{H}(\theta) \tag{12.3}$$

with $\mathcal{H}(\theta)$ the halfspace (7.46). To do this it suffices to show that if $\mathbf{x}$ belongs to the right side of (12.3), then $\mathbf{x}$ belongs to the left side. Suppose not. Then

$$\mathbf{x} \cdot \mathbf{N}(\varphi) > f(\varphi) \tag{12.4}$$

for some $\varphi \notin \Theta$. Moreover, there are adjacent preferred orientations α and β such that $0 < \beta - \alpha < \pi$ and $\varphi \in (\alpha, \beta)$. Let $\mu_\alpha(\varphi)$ and $\mu_\beta(\varphi)$ be the facet densities defined by (1.19), so that

$$f^{\#}(\varphi) = \mu_\alpha(\varphi) f(\alpha) + \mu_\beta(\varphi) f(\beta)$$

(cf. (c) of 7D, which is valid here also). Using this, the inequalities $f(\varphi) > f^{\#}(\varphi)$ and $f(\theta) \geq \mathbf{x} \cdot \mathbf{N}(\theta)$ for $\theta = \alpha, \beta$, and (1.19) (with $\mathbf{T}$ replaced by $\mathbf{N}$), we are led to $f(\varphi) > \mathbf{x} \cdot \mathbf{N}(\varphi)$, which contradicts (12.4). Thus (12.3) holds and Λ is a convex polygon.

In view of (7.48), to establish (b) and (c) it suffices to show that, given any $\Theta_0 \subset \Theta$,

$$\Lambda = \bigcap_{\theta \in \Theta_0} \mathcal{H}(\theta) \tag{12.5}$$

implies $\Theta_0 = \Theta$. Assume, to the contrary, that (12.5) holds for $\Theta_0 \subset \Theta$, $\Theta_0 \neq \Theta$. Choose $\varphi \in \Theta$, $\varphi \notin \Theta_0$. Then the line $\{\mathbf{x}: \mathbf{x} \cdot \mathbf{N}(\varphi) = f(\varphi)\}$ intersects Λ at most at a vertex of $\partial \Lambda$; hence any point $\mathbf{x}$ that satisfies

$$\mathbf{x} \cdot \mathbf{N}(\varphi) = f(\varphi) \tag{12.6}$$

must also satisfy

$$\mathbf{x} \cdot \mathbf{N}(\theta) \geq f(\theta) \qquad \text{for all } \theta \in \Theta_0. \tag{12.7}$$

Since Λ is polygonal, we may conclude from (12.5) that there are $\alpha, \beta \in \Theta_0$ such that $0 < \beta - \alpha < \pi$, $\varphi \in (\alpha, \beta)$. Then, arguing as in the proof of (a), (12.7) with $\theta = \alpha$ and $\theta = \beta$ and (12.6) imply that

$$f(\varphi) \geq \mu_\alpha(\varphi)f(\alpha) + \mu_\beta(\varphi)f(\beta).$$

Thus, recalling that the Frank diagram is the polar diagram of $f(\theta)^{-1}$, we conclude from the equivalence of (a) and (c) of (7D) that the point $\mathbf{z}$ on the Frank diagram $\mathscr{G}$ at φ lies on or inside the straight line between the vertices of $\mathscr{G}$ at α and β. Since φ is an angle of global convexity of $\mathscr{G}$, $\mathbf{z}$ cannot lie inside this line, and, by (E2), $\mathbf{z}$ cannot lie on this line. ∎

12.3 The capillary force at preferred orientations

The discussion of Chapters 4–6 remains valid provided attention is restricted to normal angles θ at which the constitutive functions are smooth. By continuity the relation

$$\sigma(\theta) = f(\theta), \qquad (12.8)$$

is valid for all θ, but since $f(\theta)$ is not differentiable at a preferred orientation α, the thermodynamic relation $(6.4)_2$, giving the shear $\xi(\theta)$ as the derivative $f'(\theta)$ of the energy, no longer has meaning at α. We will overcome this difficulty by considering ξ as *indeterminate* on each **preferred facet** (that is, on each facet whose normal angle is a preferred orientation), an assumption that leads to a capillary force of the form

$$\mathbf{C}(s, t) = f(\theta)\mathbf{T}(\theta) + \xi(s, t)\mathbf{N}(\theta) \qquad (12.9)$$

on each preferred facet, with $\xi(s, t)$ restricted only by the requirement that forces be balanced.

As before, we assume that the kinetic coefficient $b(\theta, V)$ is independent of V, strictly positive, and continuous; thus the relation

$$B(\theta, V) = -F - b(\theta)V \qquad (12.10)$$

for the normal interaction holds also on preferred facets. Since $\sigma(\theta)$ is constant on each such facet, $\sigma_s = 0$. Thus, by $(4.4)_2$, $b_{\tan} = 0$ on each preferred facet, and we may use (12.10) to write the force balance $(4.2)_1$ in the equivalent integral form

$$\int_{\partial z(t)} \mathbf{C} = \int_{z(t)} (F + b(\theta)V)\mathbf{N}(\theta)\, ds \qquad (12.11)$$

for any PS evolving curve consisting only of preferred facets. Thus, on each such facet,

$$\mathbf{C}_s = (F + b(\theta)V)\mathbf{N}(\theta), \qquad (12.12)$$

94   CRYSTALLINE ENERGIES

and, since $\theta° = \theta_s = V_s = 0$ on such facets (cf. (2.9)),

$\mathbf{C}(s, t)$ is an affine function of s on each preferred facet.   (12.13)

12.4 Corners between preferred facets

Consider a juncture i that separates facets whose normal angles are adjacent preferred orientations. Balance of forces (12.11) requires that $\mathbf{C}(s, t)$ be continuous across the juncture; thus, using the notation (2.34), $\mathbf{C}_i^-(t) = \mathbf{C}_i^+(t)$, or equivalently,

$$f(\theta_i^-)\mathbf{T}(\theta_i^-) + \xi_i^-(t)\mathbf{N}(\theta_i^-) = f(\theta_i^+)\mathbf{T}(\theta_i^+) + \xi_i^+(t)\mathbf{N}(\theta_i^+). \quad (12.14)$$

Thus, letting $\theta_i^- = \alpha$, $\theta_i^+ = \beta$ denote the adjacent preferred orientations in question, and suppressing the argument t and the subscript i, the shears $\xi^\pm$ on each side of the juncture must satisfy

$$f(\alpha)\mathbf{T}(\alpha) + \xi^-\mathbf{N}(\alpha) = f(\beta)\mathbf{T}(\beta) + \xi^+\mathbf{N}(\beta). \quad (12.15)$$

Given α and β, the equation (12.15) has a unique solution ξ^- and ξ^+. This is easily verified by taking the inner product of (12.15) with $\mathbf{N}(\alpha)$ and $\mathbf{N}(\beta)$ and writing the resulting equations as a system of equations for ξ^- and ξ^+; the determinant of this system is $1 - [\mathbf{N}(\alpha) \cdot \mathbf{N}(\beta)]^2$, which is non-zero since $|\alpha - \beta| \neq \pi$. Thus given adjacent preferred orientations α and β, there are unique shears ξ^- and ξ^+ such that (12.15) holds; the resulting vector

$$\mathbf{C}_{\text{cor}}(\alpha, \beta) = \mathbf{C}_{\text{cor}}(\beta, \alpha) \quad (12.16)$$

given by (12.15) will be referred to as the **corner force** corresponding to α and β. Thus, while the distribution of capillary forces along the facet is not known a priori, the force at the facet junctures is known; this force is, in fact, a constitutive quantity depending only on the energy and the preferred orientations in question. The next result characterizes this force in terms of the convexified Frank potential $\gamma(\theta)$ and the Wulff region Λ.

(12B) Properties of the corner force *Let α and β, with β counterclockwise of α, be adjacent preferred orientations. Then the corner force $\mathbf{C}_{\text{cor}}(\alpha, \beta)$ admits the alternative representations*

$$\mathbf{C}_{\text{cor}}(\alpha, \beta) = -f(\alpha)^2 \gamma'(\alpha + 0) = -f(\beta)^2 \gamma'(\beta - 0),$$
$$\mathbf{C}_{\text{cor}}(\alpha, \beta) = \mathbf{Q}\mathbf{x}(\alpha, \beta), \quad (12.17)$$

where the point $\mathbf{x}(\alpha, \beta)$ is the intersection of the α- and β-facets on $\partial \Lambda$, while $\mathbf{Q}$ is the clockwise rotation of $\pi/2$. Further, if μ is a preferred orientation with

$$\mathbf{C}_{\text{cor}}(\alpha, \beta) \cdot \mathbf{T}(\mu) = f(\mu), \quad (12.18)$$

then $\mu = \alpha$ or $\mu = \beta$.

CRYSTALLINE ENERGIES   95

PROOF. Let $f^{\#}(\theta)$ denote the convexified energy, and let $\mathbf{C}^{\#}(\theta)$ denote the 'capillary force' corresponding to $f^{\#}(\theta)$:

$$\mathbf{C}^{\#}(\theta) = f^{\#}(\theta)\mathbf{T}(\theta) + f^{\#\prime}(\theta)\mathbf{N}(\theta) = -f^{\#}(\theta)^2 \gamma'(\theta) \qquad (12.19)$$

(cf. (6.10), (7.11)). Then $\mathbf{C}^{\#}(\theta)$ has a jump discontinuity across each preferred orientation, but is otherwise smooth.

Since the Frank diagram of $f^{\#}(\theta)$ is a straight line between α and β, we conclude from (7D) that $\mathbf{C}^{\#}(\theta)$ is constant between these angles. Thus the limits $\mathbf{C}^{\#}(\alpha + 0)$ and $\mathbf{C}^{\#}(\beta - 0)$ coincide; writing

$$\xi^- = f^{\#\prime}(\alpha + 0), \qquad \xi^+ = f^{\#\prime}(\beta - 0), \qquad (12.20)$$

and appealing to (12.2) and (12.19), we find that

$$\mathbf{C}^{\#}(\alpha + 0) = f(\alpha)\mathbf{T}(\alpha) + \xi^- \mathbf{N}(\alpha)$$
$$= \mathbf{C}^{\#}(\beta - 0) = f(\beta)\mathbf{T}(\beta) + \xi^+ \mathbf{N}(\beta)$$
$$= -f(\alpha)^2 \gamma'(\alpha + 0) = -f(\beta)^2 \gamma'(\beta - 0). \qquad (12.21)$$

Thus the numbers $\xi^\pm$ defined by (12.20) furnish a solution of (12.15); by uniqueness, this must be *the* solution, so that (12.21) represents the corner force $\mathbf{C}_{\text{cor}}(\alpha, \beta)$. Thus (12.17)$_1$ is satisfied.

To establish (12.17)$_2$, write $\mathbf{x} = \mathbf{x}(\alpha, \beta)$. Then (12A) implies that

$$\mathbf{x} \cdot \mathbf{N}(\alpha) = f(\alpha), \qquad \mathbf{x} \cdot \mathbf{N}(\beta) = f(\beta);$$

thus there are numbers $\xi^\pm$ such that

$$\mathbf{Q}\mathbf{x} = f(\alpha)\mathbf{T}(\alpha) + \xi^- \mathbf{N}(\alpha) = f(\beta)\mathbf{T}(\beta) + \xi^+ \mathbf{N}(\beta)$$

and (12.17)$_2$ follows from the argument of the paragraph containing (12.15).

Next, by (12.17)$_2$ and (12.18),

$$\mathbf{x}(\alpha, \beta) \cdot \mathbf{N}(\mu) = f(\mu),$$

and we may conclude from (c) of (12A) that the point $\mathbf{x}(\alpha, \beta)$ lies on the line containing the μ-facet of $\partial \Lambda$, where Λ is the Wulff region. Since $\mathbf{x}(\alpha, \beta)$ also marks the intersection of the α- and β-facets of $\partial \Lambda$, and since $\partial \Lambda$ is convex, this can happen only if $\mu = \alpha$ or $\mu = \beta$. ∎

Exercise Prove the assertion containing (12.18) without using the Wulff region.

For each preferred orientation θ, let $\ell_0(\theta)$ denote the **length of the facet of $\partial \Lambda$ that has normal angle θ**, where Λ is the Wulff region.

(12C) Theorem *Let α, β, and μ denote preferred orientations with β adjacent to and counterclockwise to α, and with μ adjacent to and counterclockwise* to β. Then

$$\mathbf{C}_{\text{cor}}(\beta, \mu) - \mathbf{C}_{\text{cor}}(\alpha, \beta) = \ell_0(\beta)\mathbf{N}(\beta). \tag{12.22}$$

PROOF. Using $(12.17)_2$ and the notation of (12B),

$$\begin{aligned}\ell_0(\beta) &= \mathbf{T}(\beta) \cdot [\mathbf{x}(\alpha, \beta) - \mathbf{x}(\beta, \mu)] \\ &= -\mathbf{N}(\beta) \cdot [\mathbf{Q}\mathbf{x}(\alpha, \beta) - \mathbf{Q}\mathbf{x}(\beta, \mu)] \\ &= \mathbf{N}(\beta) \cdot [\mathbf{C}_{\text{cor}}(\beta, \mu) - \mathbf{C}_{\text{cor}}(\alpha, \beta)].\end{aligned}$$

Further, since $\mathbf{C}_{\text{cor}}(\alpha, \beta)$ is given by (12.15), and similarly for $\mathbf{C}_{\text{cor}}(\beta, \mu)$,

$$T(\beta) \cdot \mathbf{C}_{\text{cor}}(\alpha, \beta) = T(\beta) \cdot \mathbf{C}_{\text{cor}}(\beta, \mu) = f(\beta). \tag{12.23}$$

The last two results imply (12.22). ∎

12.5 Crystalline motions

We now discuss a class of evolving interfaces, termed crystalline, which are fully faceted with each facet preferred. Precisely, a **crystalline motion** is a PS evolving *interface* $\mathscr{I}$ with the following properties.

(C1) At each time the angle-set of $\mathscr{I}$ contains only preferred orientations.

(C2) Adjacent facets correspond to preferred orientations that are adjacent.

Let $\Omega(t)$ denote the reference region of $\mathscr{I}$ at each $t \in [0, T)$, with T the duration of $\mathscr{I}$; then $\Omega(t)$, $0 \le t < T$, will be termed an **evolving crystal**.

We will refer to a crystalline motion $\mathscr{I}$ (or to the corresponding evolving crystal) as **admissible** if there is a corresponding capillary force $\mathbf{C}(s, t)$ that is consistent with (12.9) and satisfies balance of forces (12.11) for every evolving subcurve $\imath$ of $\mathscr{I}$.

Let $\mathscr{I}$ be a crystalline motion. On each facet $\mathscr{I}_i$ the normal angle θ has a constant value, θ_i say, with θ_i a preferred direction. Let

$$a_i = \begin{cases} -\ell_0(\theta_i) & \text{if } \Omega \text{ is locally convex}[63] \text{ near both junctures of } \mathscr{I}_i, \\ +\ell_0(\theta_i) & \text{if } \Omega \text{ is locally concave near both junctures of } \mathscr{I}_i, \\ 0 & \text{if } \Omega \text{ is locally convex near one juncture of } \mathscr{I}_i \text{ and} \\ & \text{locally concave near the other,} \end{cases} \tag{12.24}$$

---

[63] At some (and hence every) time t. A region Γ is **locally convex** (respectively, **locally concave**) near a point $\mathbf{x} \in \partial\Gamma$ if for some sufficiently small ball $\mathscr{B}$ centred at $\mathbf{x}$, the set $\Gamma \cap \mathscr{B}$ (respectively, $[\mathbb{R}^2 \setminus \Gamma] \cap \mathscr{B}$) is convex.

or equivalently, by (2.36)$_1$,

$$a_i = \begin{cases} -\ell_0(\theta_i) & \text{if both juncture curvatures of } \mathfrak{d}_i \text{ are negative,} \\ +\ell_0(\theta_i) & \text{if both juncture curvatures of } \mathfrak{d}_i \text{ are positive,} \\ 0 & \text{if the juncture curvatures of } \mathfrak{d}_i \text{ are of opposite sign.} \end{cases} \quad (12.25)$$

We will refer to a_i as the **facet modulus** of $\mathfrak{d}_i$, and to

$$b_i = b(\theta_i).$$

as the **kinetic modulus** of $\mathfrak{d}_i$. By (12.22),

$$\mathbf{C}_{\text{cor}}(\theta_{i+1}, \theta_i) - \mathbf{C}_{\text{cor}}(\theta_i, \theta_{i-1}) = a_i \mathbf{N}(\theta_i). \quad (12.26)$$

The next result gives the counterpart, for crystalline motions, of the evolution equation $bV = gK - F$. There

$$L_i(t) = \text{the length of } \mathfrak{d}_i(t).$$

(12D) Evolution equation for crystalline motions *Necessary and sufficient that a crystalline motion be admissible is that each facet evolve according to the relation*[64]

$$b_i V_i = a_i L_i^{-1} - F. \quad (12.27)$$

PROOF. Let $\mathfrak{d}$ be an admissible crystalline motion. In view of the results of Section 12.4, the capillary force has the constant values $\mathbf{C}_{\text{cor}}(\theta_i, \theta_{i-1})$ and $\mathbf{C}_{\text{cor}}(\theta_{i+1}, \theta_i)$ at the initial and terminal junctures of $\mathfrak{d}_i$. Also, by (2.9)$_2$, on this facet $V_s = 0$ and $V = V_i = V_i(t)$. Thus (12.12) applied to $\mathfrak{d}_i(t)$ yields

$$\mathbf{N}(\theta_i) \cdot [\mathbf{C}_{\text{cor}}(\theta_{i+1}, \theta_i) - \mathbf{C}_{\text{cor}}(\theta_i, \theta_{i-1})] = [F + b(\theta_i)V_i]L_i; \quad (12.28)$$

(12.26) and (12.28) imply (12.27).

Let $\mathfrak{d}$ be a crystalline motion consistent with (12.27). To establish the admissibility of this motion we must establish the existence of a capillary force $\mathbf{C}(s, t)$ consistent with (12.9) and (12.11). Given an arbitrary facet $\mathfrak{d}_i$, we choose the capillary force $\mathbf{C}(s, t)$ on $\mathfrak{d}_i$ to have the constant values $\mathbf{C}_{\text{cor}}(\theta_i, \theta_{i-1})$ and $\mathbf{C}_{\text{cor}}(\theta_{i+1}, \theta_i)$ at the initial and terminal junctures of $\mathfrak{d}_i$, and, because of (12.13), we define $\mathbf{C}(s, t)$ to be the affine function (in s) that connects these constant values. Then, by (12.23), $\mathbf{C}(s, t) \cdot \mathbf{T}(\theta_i) = f(\theta_i)$ on $\mathfrak{d}_i$; hence $\mathbf{C}(s, t)$ is consistent with (12.9). Also,

$$\mathbf{C}_s = L_i^{-1}[\mathbf{C}_{\text{cor}}(\theta_{i+1}, \theta_i) - \mathbf{C}_{\text{cor}}(\theta_i, \theta_{i-1})]$$

---

[64] Taylor (1988) proposed (12.27) for the special case $F = 0$, $b_i = f(\theta_i)^{-1}$ (cf. (12G) and Taylor (1991)). The general equation (12.27) was arrived at independently by Angenent and Gurtin (1989).

on ∂_i, so that, by (12.26) and (12.27), **C** obeys (12.12) on ∂_i. Thus, since i is an arbitrary juncture, and since $\mathbf{C}(s, t)$ is continuous in s across each juncture, balance of forces (12.11) is satisfied. ∎

Equations (12.27) and (2.46) form the following result.

(12E) Ordinary differential equations for crystalline motions *Admissible crystalline motions evolve according to the following system of ordinary differential equations*:

$$L_i^{\cdot} = (\cot \vartheta_{i+1} + \cot \vartheta_i) V_i - (\sin \vartheta_i)^{-1} V_{i-1} - (\sin \vartheta_{i+1})^{-1} V_{i+1},$$

$$V_i = \frac{1}{b_i}\left[-F + \frac{a_i}{L_i}\right] \quad (12.29)$$

for each juncture i.

Curves governed by smooth energies have, as evolution equations, a fairly complex system of partial differential equations; the theorem above shows that, in contrast, crystals corresponding to non-smooth energies (that are crystalline) evolve according to a finite set of ordinary differential equations, one equation for each facet.

(12F) Remarks

1. The system (12.29) determines a crystalline motion from an initial closed curve that is fully faceted and consistent with (C1) and (C2); indeed, the normal angles θ_i, the juncture curvatures ϑ_i, the constants (12.26), and the initial lengths $L_i(0)$ can be determined from the initial curve.

2. Knowledge of the lengths $L_i(t)$ and the preferred orientations characterizes the crystal at t only up to a rigid translation. However the trajectory $\mathbf{R}_i(t)$ of any juncture i may be found using the relation

$$\mathbf{R}_i^{\cdot}(t) = V_{i+1}\mathbf{N}_{i+1} + (\sin \vartheta_i)^{-1}\{V_i - (\cos \vartheta_i)V_{i+1}\}\mathbf{T}_{i+1}. \quad (12.30)$$

3. A theorem analogous to (10D) (the theorem on the growth of the reference phase) is valid for evolving crystals.[65]

Exercise Derive (12.30).

Consider an evolving crystal $\Omega(t)$ that is convex at some (and hence every) t. It follows from our sign conventions (cf. Sections 1.1 and 2.1) that as we traverse the curve in the direction of increasing arc length, or equivalently, as we traverse the junctures i in their numerical order, the normal angles θ_i

---

[65] Cf. Angenent and Gurtin (1989), p. 380.

decrease; thus letting N denote the essential number of facets, we may order the evolving arcs so that

$$0 \leq \theta_N < \theta_{N-1} < \cdots < \theta_1 < 2\pi. \tag{12.31}$$

It then follows from properties (C1) and (C2) of crystalline motions that the list (12.31) also represents the complete list of vertex angles of the (polygonal) convexified Frank diagram.

Soner's explicit solution (10K) for smooth energies has an analogue for crystalline energies.

(12G) Taylor's solution[66] *Consider the special case in which*

$$b(\theta)f(\theta) = \kappa \; (=\text{constant}) \tag{12.32}$$

for all preferred orientations θ. Let

$$\Omega(t) = z(t)\Lambda,$$

$z(t) > 0$, $0 \leq t < T$, so that $\Omega(t)$ is a dilation of the Wulff region Λ (cf. (7.42)). Then $\Omega(t)$, $0 \leq t < T$, is an admissible evolving crystal if and only if $z(t)$ is a solution of the differential equation

$$\kappa z^{\cdot}(t) = -[F + z(t)^{-1}]. \tag{12.33}$$

PROOF. Letting (12.31) denote the preferred orientations, the length of the θ_i-facet of $\partial\Omega(t)$ is given by $L_i(t) = z(t)\ell_0(\theta_i)$. Let $p_i(t)$ denote the distance, from the origin, of the θ_i-facet of $\partial\Omega(t)$. By (c) of (12A), this distance for $\partial\Lambda$ is $f(\theta_i)$. Thus $p_i(t) = z(t)f(\theta_i)$, so that $V_i(t) = z^{\cdot}(t)f(\theta_i)$. Thus, since $\Omega(t)$ is convex, we may conclude from (12.24), (12.27), and (12.32) that $\Omega(t)$ is an admissible evolving crystal if and only if (12.33) is satisfied. ∎

The next result compares crystal growth for two sets of data: $f(\theta)$, $b(\theta)$, F and $\underline{f}(\theta)$, $\underline{b}(\theta)$, $\underline{F}$. The energies f and $\underline{f}$ are assumed to be crystalline, with the same set of preferred orientations, so that we may write $\ell(\varphi)$ and $\underline{\ell}(\varphi)$, respectively, for the lengths of facets of normal angle φ on the Wulff regions corresponding to f and $\underline{f}$.

(12H) Containment principle *Assume that, for each preferred orientation φ,*

$$b(\varphi)^{-1}\ell(\varphi) \geq \underline{b}(\varphi)^{-1}\underline{\ell}(\varphi), \qquad Fb(\varphi) \leq \underline{F}\underline{b}(\varphi).$$

Then given admissible evolving crystals $\Omega(t)$ and $\underline{\Omega}(t)$, $0 \leq t < T$, for the two systems,

$$\Omega(0) \subset \underline{\Omega}(0) \quad \text{implies} \quad \Omega(t) \subset \underline{\Omega}(t), \quad 0 \leq t < T.$$

Exercise Prove this result.

---

[66] Taylor (1988) for the case $F = 0$.

12.6 Interfaces of arbitrary orientation, infinitesimal wrinklings, and generalized motions

Thus far we have restricted attention to interfaces whose normal angles are preferred orientations; such interfaces are necessarily fully faceted. We now discuss the evolution—for a crystalline energy—of an interface with arbitrary orientation. As before we extend the theory to arbitrary orientations by allowing the interface to form infinitesimal wrinkles between preferred orientations.

Consider a wrinkling ω that is an evolving subcurve of an admissible crystalline motion $\mathfrak{s}$. Let ω have normal angles α and β, with α and β adjacent preferred orientations, and assume that each of the two facets of $\mathfrak{s}$ that connect to ω has normal angle α or β. Then admissibility requires that $\mathbf{C}(s, t)$ have the constant value $\mathbf{C}_{\text{cor}}(\alpha, \beta)$ at the initial and terminal point of ω and at each of the junctures of $\mathfrak{s}$; thus, by (12.13),

$$\mathbf{C}(s, t) \equiv \mathbf{C}_{\text{cor}}(\alpha, \beta) \qquad \text{on } \omega. \tag{12.34}$$

Further, by (12.24), $a_i = 0$ for each facet of ω, so that, by (12.27), the normal velocity V satisfies (a) and hence (c) of (11I). Thus *wrinklings have essentially the same properties as they did for smooth energies*, and, as for such energies, we will adopt these conditions as defining relations for an infinitesimal wrinkling.

An **admissible infinitesimal wrinkling** with normal angles α and β is a PS evolving curve ω that satisfies (W1) and (W2) of Section 11.5 with α and β adjacent preferred orientations. The **corresponding capillary force** is then defined to have the constant value (12.34). It then follows that ω evolves according to the evolution equation (11.22), with $b_{\text{eff}}(\theta)$, the **effective kinetic modulus**, defined as in the paragraph containing (11.21). Note that, because the energy is crystalline, the polar diagram of $b_{\text{eff}}(\theta)$ is polygonal with vertices $b(\varphi)\mathbf{N}(\varphi)$ at preferred orientations φ.

We now use the notion of an infinitesimal wrinkling to generalize the concept of a crystalline motion. Let $\mathfrak{s}$ be a PS evolving *interface* that is the union of

PS evolving curves c_p, $p = 1, \ldots, P$, consistent with (C1) and (C2),

and

admissible infinitesimal wrinklings ω_p, $p = 1, \ldots, P$.

We will refer to the curves c_p as **evolving crystalline curves**; we will use the term **subfacet** of $\mathfrak{s}$ to designate a facet of any one of the c_p's; we will use the term **subwrinkling** of $\mathfrak{s}$ to denote any one of the ω_p's. Then $\mathfrak{s}$ is a **generalized, admissible crystalline motion** if the following conditions are satisfied.

(G1) The evolving crystalline curves alternate with the subwrinklings.

(G2) Given any c_p, there is a corresponding capillary force $\mathbf{C}(s, t)$ on c_p that is consistent with (12.9), satisfies balance of forces (12.11) for every evolving subcurve $\imath$ of c_p, and, at the initial and terminal points of c_p, equals the capillary force on the adjoining subwrinkling.

We then have the following analogue of (W3).

(G3) Given a subfacet $\imath$ that adjoins a ω_p, the normal angle of $\imath$ must be one of the two normal angles of ω_p.

To verify (G3), let α and β denote the normal angles of ω_p, and let μ denote the normal angle of $\imath$. Then, by (12.34), the capillary force on ω_p is the constant $\mathbf{C}_{\text{cor}}(\alpha, \beta)$, so that, by (G2), the capillary force on $\imath$ must have this value at its juncture with ω_p. Thus, by (12.9), $\mathbf{C}_{\text{cor}}(\alpha, \beta) \cdot \mathbf{T}(\mu) = f(\mu)$, and we conclude from the assertion containing (12.18) that $\mu = \alpha$ or $\mu = \beta$, which is (G3).

Given a subfacet $\imath$ of ∂, we write $a(\imath)$ for the *facet modulus* of $\imath$ as defined by (12.25); $a(\imath)$ depends on $\imath$ only through the normal angle of $\imath$ and the convexity of its junctions, and may be calculated from the initial data.

(12I) Evolution equation for generalized crystalline motions *Generalized crystalline motions ∂ evolve as follows. Each subfacet $\imath$ of ∂ evolves according to*

$$b_{\text{eff}}(\theta)V = a(\imath)L^{-1} - F \qquad (12.35)$$

with $L(t)$ the length of $\imath(t)$; each subwrinkling ω of ∂ evolves according to

$$b_{\text{eff}}(\theta)V = -F. \qquad (12.36)$$

PROOF. We have already noted that (12.36) holds on infinitesimal wrinklings. To verify that (12.35) is satisfied, let $\imath$ be an arbitrary subfacet of ∂. Then the two neighbours of $\imath$ are either subfacets, or a subfacet and a subwrinkling, or two subwrinklings. If the neighbours are subfacets, then the analysis leading to (12D) remains valid, so that (12.35) is satisfied.

Suppose that $\imath$ has, as a neighbour, a subwrinkling ω, and let j denote the corresponding juncture. Then, by (G3), the normal angle α of $\imath$ must be one of the two normal angles of ω. Let the other normal angle be β.

Assertion The sign of the juncture curvature at j and the capillary force on $\imath$ at j are the same as they would be if the subwrinkling were replaced by a *facet* of normal angle β.

PROOF (Assertion). By (12.34), the capillary force on ω is $\mathbf{C}_{\text{cor}}(\alpha, \beta)$, so that, by (G2), the force on $\imath$ at j must also be $\mathbf{C}_{\text{cor}}(\alpha, \beta)$; this is what the force would be if ω were replaced by a facet of normal angle β. We will prove the remaining statement only for the case in which j coincides with the terminal point of $\imath$ and $0 < \beta - \alpha < \pi$; the proof for each of the remaining cases is similar. Since the angle-set of ω is contained in (α, β), the juncture curvature at j is >0 (cf. $(2.36)_1$), which is the same as it would be if ω were replaced by a facet of normal angle β. ∎

In view of the assertion, the verification of (12.35) for the cases in which one or both of the neighbours of $\imath$ are wrinklings again follows using the analysis leading to (12D). This completes the proof of (12I). ∎

(12J) Decay theorem for infinitesimal wrinklings *Let ω be a subwrinkling of a generalized, admissible crystalline motion. Then*

$$\omega(t_2) \subset \omega(t_1) + (t_2 - t_1)\mathbf{w} \qquad \text{for } t_2 > t_1, \qquad (12.37)$$

so that, modulo a rigid translation, $\omega(t)$ nests as t increases. Further, the inclusion in (12.37) is strict if the juncture curvatures, of any one of the two subfacets that adjoin ω, are of the same sign.

PROOF. The proof is almost identical to that of (11L). Let $\imath_C$ and $\imath_D$, respectively, denote the subfacets that adjoin ω at its initial and terminal points. Consider the juncture between $\imath_C$ and ω. Let θ_c, $V_c(t)$, $v_c(t)$, a_c, $\mathbf{T}_c$, $\mathbf{N}_c$, respectively, denote the normal angle, normal velocity, tangential endpoint velocity, facet modulus, tangent, and normal for $\imath_C$ at this juncture, let $\theta_w(t)$, $V_w(t)$, $v_w(t)$, $\mathbf{T}_w(t)$, $\mathbf{N}_w(t)$ denote corresponding quantities for ω, and let $\mathbf{w}$ denote the rigid velocity of ω. To establish the result (12.37) it suffices to show that

$$v_w(t) \geq \mathbf{w} \cdot \mathbf{T}_w(t), \qquad (12.38)$$

that the inequality is strict if both junctures of $\imath_C$ have the same sign, and that the relative tangential velocity of the juncture between ω and $\imath_D$ satisfies analogous inequalities with $\geq$ and $>$ replaced by $\leq$ and $<$. As before we will establish only those results concerning (12.38).

The steps leading to (11.25) here yield

$$\begin{aligned} v_w - \mathbf{w} \cdot \mathbf{T}_w &= (\sin \vartheta)^{-1}\{V_c - \mathbf{w} \cdot \mathbf{N}_c\} \\ &= (\sin \vartheta)^{-1}\{V_c - \mathbf{w} \cdot \mathbf{N}_c\} \\ &= (\sin \vartheta)^{-1} a_c/L_c \end{aligned} \qquad (12.39)$$

with $\vartheta \neq 0$ the juncture curvature between $\imath_C$ and ω. Let α and β, $0 < \beta - \alpha < \pi$, denote the normal angles of ω. By (W3), the angle θ_c is either α or β; assume $\theta_c = \alpha$. Then, by (W1) and $(2.36)_1$, $\vartheta > 0$. Thus, by (12.25), $a_c \geq 0$, so that (12.39) is ≥ 0, with the inequality strict if $a_c > 0$, which results if both juncture curvatures of $\imath_C$ have the same sign. If $\theta_c = \beta$, then $\vartheta < 0$, $a_c \leq 0$, and we have the same conclusion regarding (12.39). ∎

12.7 Evolution of a rectangular crystal

A simple example that yields useful information occurs when the convexified Frank diagram is quadrilateral with vertices at $\theta = 0$, $\pi/2$, π, $3\pi/2$. A crystal of a corresponding admissible crystalline motion, if convex, is then rectangular at each t with sides having these angles as orientations. Thus $L_1 = L_3$, $L_2 = L_4$, and, defining

$$F_1 = \frac{F(b_2 + b_4)}{b_2 b_4}, \qquad F_2 = \frac{F(b_1 + b_3)}{b_1 b_3},$$
$$\delta_1 = -\frac{a_4 b_2 + a_2 b_4}{b_2 b_4}, \qquad \delta_2 = -\frac{a_3 b_1 + a_1 b_3}{b_1 b_3}, \tag{12.40}$$

the evolution equations (12.29) reduce to

$$L_1^{\cdot} = -F_1 - \frac{\delta_1}{L_2},$$
$$L_2^{\cdot} = -F_2 - \frac{\delta_2}{L_1} \tag{12.41}$$

with (cf. (12.24))

$$\operatorname{sgn} F_i = \operatorname{sgn} F, \qquad \delta_i > 0. \tag{12.42}$$

Case 1: $F = 0$. Solutions of (12.41) approach zero in *finite* time t_∞. Defining $\delta = \delta_1/\delta_2$, there is a constant $C > 0$ such that

$$L_1(t) = C L_2(t)^\delta. \tag{12.43}$$

For $\delta = 1$ the *isoperimetric ratio*

$$\rho = \frac{1}{4\pi} \frac{\operatorname{length}(\partial \Omega)^2}{\operatorname{area}(\Omega)}$$

is constant, but for $\delta \neq 1$

$$\rho(t) \to \infty \qquad \text{as } t \to t_\infty. \tag{12.44}$$

For $\delta > 1$, $L_1(t)$ approaches zero faster than $L_2(t)$, so that the crystal shrinks to a point, but is ultimately in the shape of a 'needle oriented by θ_2 and θ_4'. This is in sharp contrast to an evolving interface that is isotropic (cf. Section 8.1); there the interface shrinks to a round point.[67]

Case 2: $F > 0$. Solutions still approach zero in finite time t_∞. The result (12.43) holds asymptotically, and the discussion of Case 1 for $\delta \neq 1$ is appropriate.

---

[67] Gage (1984); Gage and Hamilton (1986); Grayson (1987).

Case 3: $F < 0$. This case corresponds to a crystal evolving in a supercooled liquid. Here (12.41) has an equilibrium at

$$L_2 = \frac{|F_1|}{\delta_1}, \qquad L_1 = \frac{|F_2|}{\delta_2}, \qquad (12.45)$$

which is a saddle. For any given initial value $L_1(0)$ there is a number $\ell > 0$ such that: (i) if $L_2(0) < \ell$, the sides shrink to zero in finite time, in which case the asymptotic behaviour of the crystal is as discussed in Case 1; and (ii) if $L_2(0) > \ell$, the sides grow to infinity as $t \to \infty$, asymptotically as

$$L_1(t) \approx |F_1|t, \qquad L_2(t) \approx |F_2|t.$$

13
REGULARIZED THEORY FOR SMOOTH UNSTABLE ENERGIES; DEPENDENCE OF INTERFACIAL ENERGY ON CURVATURE[68]

Unstable interfacial energies lead to backward-parabolic evolution equations for the interface. In this section we regularize the evolution equations by allowing for a curvature-dependent interfacial energy, a dependence that requires the introduction of capillary moments in conjunction with a more general balance law for moments.

13.1 Balance of forces and moments; power

Consider an evolving interface $\mathscr{s}$ with $\mathbf{r}$ a corresponding parametrization. The micromechanics of the interface is described by four interfacial fields: the capillary force $\mathbf{C}(s, t)$, the interactive force $B(s, t)$, and the interactive moment $m(s, t)$, discussed in Section 4.1, and an additional field, the **capillary moment** $M(s, t)$, which represents the moment *within the interface* exerted across s.

Balance of forces and moments for an evolving subcurve $\imath$ (interpreted as an infinitesimally thin two-phase evolving control volume) is the requirement that

$$\int_{\partial \imath} \mathbf{C} + \int_{\imath} \mathbf{B} \, ds = \mathbf{0},$$

$$\int_{\partial \imath} (\mathbf{r} \times \mathbf{C} + M) + \int_{\imath} (\mathbf{r} \times \mathbf{B} + m) \, ds = 0. \tag{13.1}$$

These laws must hold for all interfacial subcurves $\imath(t)$, and hence have local forms

$$\mathbf{C}_s + \mathbf{B} = \mathbf{0}, \qquad M_s + m + \xi = 0 \tag{13.2}$$

(cf. (4.3), (4.4)). As before, what is important is the normal component of $(13.2)_1$,

$$\xi_s + \sigma K + B = 0, \tag{13.3}$$

---
[68] Cf. Di Carlo et al. (1992).

and this with (13.2)$_2$ yields the basic balance law of the theory:

$$M_{ss} + m_s - \sigma K - B = 0. \tag{13.4}$$

In these relations we use the decomposition (4.1) of the capillary force into a surface tension σ and a surface shear ξ, and we define the normal interaction B through (5.9).

The forces **C** and **B** and the moment m expend power as described in (4.6) and (4.7). The moment M acts on the endpoints of $\imath(t)$; we assume that M expends power over the rotation rate of the interface following the endpoints. Denoting by $[S_1(t), S_2(t)]$ the interval of arc lengths for $\imath(t)$, and by $\Theta_i(t) = \theta(S_i(t), t)$ the angle of the endpoint $S_i(t)$, we define the rotation rate $(\theta_{\partial \imath})^{\cdot}(s, t)$ at $s = S_1(t)$ and $s = S_2(t)$ through

$$(\theta_{\partial \imath})^{\cdot}(s, t) = \Theta_i^{\cdot}(t). \tag{13.5}$$

The *power expended* on $\imath(t)$ is then given by

$$\int_{\partial \imath} [\mathbf{C} \cdot \mathbf{v}_{\partial \imath} + M(\theta_{\partial \imath})^{\cdot}] + \int_{\imath} (\mathbf{B} \cdot \mathbf{V} \mathbf{N} + m\theta^{\circ}) \, ds, \tag{13.6}$$

and, by (2.9)$_{2,3}$, (2.16)$_2$, (13.2)$_2$, and (13.3), the power expense (13.6) is equal to

$$\int_{\imath} [MK^{\circ} - (\sigma + MK)KV] \, ds + \int_{\partial \imath} (\sigma + MK) v_{\partial \imath (\tan)} \tag{13.7}$$

(cf. (4.9)). The term MK° represents power expended in bending the interface, while $-(\sigma + MK)KV$ represents power expended in creating new interface. Here, interestingly, *both the surface tension and the capillary moment work to create new surface*.

13.2 Energetics and the dissipation inequality

With each interfacial motion we associate an interfacial energy $f(s, t)$ in conjunction with a constant F that represents the difference in bulk energies (cf. (5.1), (5.10)); and we postulate an **interfacial dissipation inequality** of the form

$$\frac{d}{dt} \int_{\imath} f \, ds + \int_{\imath} FV \, da \leq \int_{\partial \imath} [\mathbf{C} \cdot \mathbf{v}_{\partial \imath} + M(\theta_{\partial \imath})^{\cdot}], \tag{13.8}$$

containing a term representing power expended by the capillary moment.

The dissipation inequality (13.8) is required to hold for all evolving subcurves $\imath$ of $\jmath$. Thus, proceeding as in the proof of (5.7) and (5.8), but using

REGULARIZED THEORY FOR SMOOTH UNSTABLE ENERGIES 107

the equivalence of (13.6) and (13.7), we are led to the inequality

$$\int_{\imath} \{f^\circ - MK^\circ + m\theta^\circ - (f - \sigma - MK)KV + (B + F)V\} \, ds$$

$$+ \int_{\partial \imath} (f - \sigma - MK)\mathfrak{v}_{\partial\imath(\tan)} \leq 0, \quad (13.9)$$

and then to two important results: (i) the identity

$$f = \sigma + MK, \quad (13.10)$$

showing that *capillary moments negate the identification of surface tension with interfacial energy*; and (ii) the *reduced dissipation inequality*

$$f^\circ - MK^\circ + m\theta^\circ + (B + F)V \leq 0. \quad (13.11)$$

(13A) Remark We consider the tangential force $B_{\tan}$ and the surface shear ξ as *indeterminate*,[69] since neither of these quantities enters the dissipation inequality (13.11). By (4.4)$_3$, this does not contradict our previous treatment in which m rather than ξ was chosen to be indeterminate.

13.3 Constitutive equations

We begin with *smooth* constitutive equations

$$f = f(\theta, K, V), \qquad M = M(\theta, K, V),$$
$$m = m(\theta, K, V), \qquad B = B(\theta, K, V), \quad (13.12)$$

thereby allowing the curvature K to join the variables effecting the energy. Because of (13.10) we do not specify a separate constitutive equation for the surface tension σ, and we do not write constitutive equations for the indeterminate forces ξ and $B_{\tan}$. In principle, once the balance equation (13.4) has been solved for the interfacial motion, then (13.3) determines ξ and, with this, the tangential component of (13.2)$_1$ determines $B_{\tan}$.

Given an evolving interface $\mathfrak{s}$, the constitutive equations may be used to compute a corresponding *constitutive process* (f, M, B, m). If all constitutive processes are consistent with the reduced dissipation inequality (13.11), we will refer to the constitutive equations as **compatible with thermodynamics**.

---

[69] This is consistent with classical beam theory, in which the shear force—regarded as a reaction to the constraint that the cross-section remain normal to the centre line—is indeterminate.

Substituting (13.12) into (13.11), we find that

$$f_V(\theta, K, V)V^\circ + [f_K(\theta, K, V) - M(\theta, K, V)]K^\circ$$
$$+ [f_\theta(\theta, K, V) + m(\theta, K, V)]\theta^\circ + [B(\theta, K, V) + F]V \le 0,$$

and arguing as in the proof of (6A), we arrive at the next result.

(13B) Compatibility theorem *The constitutive equations are compatible with thermodynamics if and only if*:

(a) *the energy, capillary moment, and interactive moment are independent of V and satisfy*

$$m(\theta, K) = -f_\theta(\theta, K), \qquad M(\theta, K) = f_K(\theta, K); \qquad (13.13)$$

(b) *the normal interaction has the form*

$$B(\theta, K, V) = -F - b(\theta, K, V)V,$$
$$b(\theta, K, V) \ge 0. \qquad (13.14)$$

(13C) Remarks

1. The restrictions (13.13) yield the Gibbs relation

$$f^\circ = MK^\circ - m\theta^\circ. \qquad (13.15)$$

2. By $(13.13)_2$, capillary moments are necessary for a dependence of interfacial energy on curvature.

3. In view of (13.14) and (13.15), the left side of (13.11) is $-b(\theta, K, V)V^2$, which identifies this quantity as the *sole* rate of energy dissipation; indeed, tracing backwards the argument leading to (13.11), we find that

$$-\int_s b(\theta, K, V)V^2 \, ds \qquad (13.16)$$

represents the left side of (13.8) minus the right, and this, in turn, leads to the following counterpart of the growth theorem (10.7) for the reference region $\Omega(t)$ of an evolving interface $s(t) = \partial\Omega(t)$:

$$\frac{d}{dt}\left\{\int_{\partial\Omega} f(\theta, K) \, ds + F \, \text{area}(\Omega(t))\right\} = -\int_{\partial\Omega} b(\theta, K, V)V^2 \, ds \le 0.$$

$$(13.17)$$

Exercise Establish (13.9)–(13.11) and (13.17), and prove the compatibility theorem.

13.4 Evolution equations for the interface

The resulting evolution equation is (13.4) supplemented by (13.10), (13.13), and (13.14):

$$M_{ss} + m_s - fK + MK^2 + F + bV = 0,$$
$$f = f(\theta, K), \qquad m = -f_\theta(\theta, K),$$
$$M = f_K(\theta, K), \qquad b = b(\theta, K, V). \tag{13.18}$$

This general equation appears difficult, and for that reason we consider a model in which f has a relatively simple dependence on K, and b depends only on θ:

$$f = f_0(\theta) + \tfrac{1}{2}\varepsilon K^2, \qquad b = b_0(\theta), \tag{13.19}$$

with $f_0(\theta), b_0(\theta), \varepsilon > 0$. Then

$$m = -f'_0(\theta), \qquad M = \varepsilon K, \tag{13.20}$$

and, writing $b(\theta) = b_0(\theta)$, (13.18) reduces to[70]

$$b(\theta)V = g(\theta)K - \varepsilon(K_{ss} + \tfrac{1}{2}K^3) - F,$$
$$g(\theta) = f_0(\theta) + f''_0(\theta), \tag{13.21}$$

which is the anisotropic evolution equation

$$b(\theta)V = g(\theta)K - F \tag{13.22}$$

regularized by the higher-order term $\varepsilon(K_{ss} + \tfrac{1}{2}K^3)$.

(13D) Remark Chapter 11 discusses (13.22) using admissible corners to exclude angles at which the equation exhibits backward-parabolic behaviour. The regularized equation (13.21) should be appropriate to study the behaviour of the interface within such corners. If θ_1 and θ_2 with $0 < \theta_2 - \theta_1 < \pi$ are the angles that define an admissible corner, then the behaviour of the interface within the corner might be modelled by the system

$$b(\theta)V = g(\theta)K - \varepsilon(K_{ss} + \tfrac{1}{2}K^3) - F \qquad -\infty < s < +\infty, \ t > 0,$$
$$\theta(-\infty, t) = \theta_1, \quad \theta(+\infty, t) = \theta_2 \qquad t > 0,$$
$$\theta(s, 0) = \theta_0(s) \qquad -\infty < s < +\infty,$$

with initial data $\theta_0(s) \in (\theta_1, \theta_2)$.

For a *convex interface* the relevant differential equations are (2.20) and

---

[70] At present there are neither analytical nor numerical studies concerning this equation.

(13.21), or equivalently, since $K_{ss} = \frac{1}{2}K(K^2)_{\theta\theta}$,

$$K_t = K^2(V_{\theta\theta} + V),$$
$$b(\theta)V = g(\theta)K - \tfrac{1}{2}\varepsilon K[(K^2)_{\theta\theta} + K^2] - F \tag{13.23}$$

(with K_t the derivative holding θ fixed).

13.5 Linearized equations; spinodal decomposition on the interface

We now consider an interface that is close to a flat interface at angle $\theta = \pi/2$. We represent the interface as the graph of a function $y = h(x, t)$ and use the conventions of Section 8.2.2 and the relations (8.10), (8.11), and $\partial/\partial s = (\sin \theta) \, \partial/\partial x$ to linearize (13.21) under the assumption that the derivatives of h are small; the result is

$$h_t = \alpha h_{xx} - \delta h_{xxxx} - \beta, \tag{13.24}$$

where $\alpha = g_0/b_0$, $\beta = F/b_0$, $\delta = \varepsilon/b_0$, and the subscript zero here indicates evaluation at $\theta = \pi/2$.

Differentiating (13.24) with respect to x yields

$$w_t = \alpha w_{xx} - \delta w_{xxxx}, \tag{13.25}$$

with $w = h_x$, which is the linear equation used by Cahn (1968) in his discussion of spinodal decomposition.[71] Following Cahn, we note that (13.25) has solutions of the form

$$w(x, t) = C \, e^{\rho t} \sin \lambda x \tag{13.26}$$

provided

$$\rho = \rho(\lambda) = -\alpha \lambda^2 - \delta \lambda^4. \tag{13.27}$$

If $g(\pi/2) < 0$, then $\alpha < 0$ and spatial oscillations of frequency λ between 0 and $(|\alpha|/\delta)^{\frac{1}{2}}$ are unstable, with a maximum of $\rho(\lambda)$ occurring at $\lambda = \lambda_m = (\frac{1}{2}|\alpha|/\delta)^{\frac{1}{2}}$; this indicates that spatial patterns of frequency λ_m should be most often observed.

---

[71] In fact, the angle intervals in which $g(\theta) < 0$ are very much like the spinodals encountered in the Korteweg–Cahn–Hilliard theory of phase transitions, while the globally stable sections of $f_0(\theta)$ are analogous to standard bulk phases.

III
THERMODYNAMICAL THEORY OF INTERFACIAL EVOLUTION IN THE PRESENCE OF BULK HEAT CONDUCTION

In the theory presented thus far the bulk material manifests itself through a constant difference in bulk energies: transport processes such as the diffusion of heat and mass in bulk are neglected. In this part we discuss interfacial evolution in the presence of bulk heat conduction; as we shall see, non-uniform bulk behaviour can have a profound influence on the evolution of the interface.

To introduce the basic thermodynamic concepts, we begin with a discussion of single-phase materials.

14
REVIEW OF SINGLE-PHASE THERMODYNAMICS[72]

14.1 Basic quantities and the first two laws

We assume that the body occupies a fixed region Ω in $\mathbb{R}^2$, and that the material is immobile, incapable of deformation, with heat conduction the sole transport mechanism. We base the theory on *dynamical* versions of the first two laws applied to arbitrary control volumes (subregions R of Ω).

> The *first law, balance of energy*, asserts that the internal
> energy of R change at a rate balanced by the heat flow into
> R and the power expended on R. The *second law, growth of
> entropy*, requires that the entropy of R change at a rate
> not less than the entropy flow into R. (14.1)

For this *single-phase* theory there is no expense of power, since the bulk material is immobile, and since there is no phase interface.

The thermodynamics of Ω is described by four **bulk fields** defined on Ω for all time:

$T(\mathbf{x}, t)$ absolute temperature,

$\varepsilon(\mathbf{x}, t)$ bulk internal energy,

$\eta(\mathbf{x}, t)$ bulk entropy,

$\mathbf{q}(\mathbf{x}, t)$ bulk heat flux,

$r(\mathbf{x}, t)$ bulk heat supply,

with ε, η, and $T > 0$ scalar fields and $\mathbf{q}$ a vector field.

Let R be a control volume, with $\mathbf{n}$ the outward unit normal to ∂R. Then

$$\int_R \varepsilon \, da$$

represents the internal energy of R;

$$-\int_{\partial R} \mathbf{q} \cdot \mathbf{n} \, ds \qquad (14.2)$$

gives the net heat flow into R by conduction across ∂R;

$$\int_R r \, da$$

---

[72] Cf. Coleman and Noll (1963); Coleman and Mizel (1963).

gives the total heat supplied to R by the external world. The precise statement of the **first law (balance of energy)** for R is then

$$\frac{d}{dt}\int_R \varepsilon \, da = -\int_{\partial R} \mathbf{q}\cdot\mathbf{n}\, ds + \int_R r \, da. \qquad (14.3)$$

The integral

$$\int_R \eta \, da$$

represents the internal entropy of R. Each density Q of heat flow induces a corresponding density Q/T of entropy flow; in particular, the conductive heat flow $-\mathbf{q}\cdot\mathbf{n}$ across ∂R and the heat r supplied to the points of R induce entropy-flow densities $-\mathbf{q}\cdot\mathbf{n}/T$ and r/T, so that

$$-\int_{\partial R}\left(\frac{\mathbf{q}}{T}\right)\cdot\mathbf{n}\, ds, \quad \int_R \frac{r}{T}\, da \qquad (14.4)$$

represent corresponding net entropy flows into R. The precise statement of the **second law (growth of entropy)** for R is then

$$\frac{d}{dt}\int_R \eta \, da \geq -\int_{\partial R}\left(\frac{\mathbf{q}}{T}\right)\cdot\mathbf{n}\, ds + \int_R \frac{r}{T}\, da. \qquad (14.5)$$

Since the control volume R is a fixed region, the time derivatives in (14.3) and (14.5) may be taken under the integral; thus, applying the divergence theorem to the right sides of these equations, and using the fact that R is arbitrary, we are led to the local energy balance

$$\dot{\varepsilon} = -\operatorname{div}\mathbf{q} + r \qquad (14.6)$$

and the the local law of entropy growth

$$\dot{\eta} \geq -\operatorname{div}\left(\frac{\mathbf{q}}{T}\right) + \frac{r}{T}. \qquad (14.7)$$

The field

$$\Gamma = \dot{\eta} + \operatorname{div}\left(\frac{\mathbf{q}}{T}\right) - \frac{r}{T} \geq 0 \qquad (14.8)$$

represents the **bulk entropy production**, since, trivially,

$$\frac{d}{dt}\int_R \eta \, da + \int_{\partial R}\left(\frac{\mathbf{q}}{T}\right)\cdot\mathbf{n}\, ds - \int_R \frac{r}{T}\, da = \int_R \Gamma \, da \qquad (14.9)$$

for any control volume R.

A useful local version of the second law involves the **bulk free energy**

$$\psi = \varepsilon - T\eta \qquad (14.10)$$

and is obtained by eliminating div $\mathbf{q}$ between (14.7) and (14.8); the result is the **bulk free-energy inequality**:

$$\psi^{\cdot} + \eta T^{\cdot} + T^{-1}\mathbf{q}\cdot\text{grad } T = -T\Gamma \leq 0. \qquad (14.11)$$

An immediate consequence of (14.11) is that *the free energy cannot increase in an isothermal process* (a process with $T \equiv$ constant). As we shall see, the free energy is important also in non-isothermal processes, especially when discussing constitutive equations with temperature as an independent variable.

Exercise Show that if $r = 0$ and if $T(\mathbf{x}, t) = T_0(t)$ on a portion of $\partial\Omega$ and $\mathbf{q}(\mathbf{x}, t)\cdot\mathbf{n}(\mathbf{x}) = 0$ on the remainder, with $\mathbf{n}$ the outward unit normal to $\partial\Omega$, then

$$\frac{d}{dt}\int_\Omega (\varepsilon - T_0\eta)\, da = \frac{d}{dt}\int_\Omega [\psi + (T - T_0)\eta]\, da \leq 0.$$

14.2 Constitutive equations and thermodynamic restrictions

The simplest law of heat conduction is that proposed by Fourier giving the heat flux $\mathbf{q}$ as an isotropic, linear function of the temperature gradient:

$$\mathbf{q} = -k\,\text{grad } T. \qquad (14.12)$$

The constant k is called the *conductivity* and is taken to be strictly positive to ensure that heat flows from hot to cold. With (14.12) one classically adjoins 'equations of state' giving, for example, the internal energy and the entropy in terms of the temperature. Here we shall take a more general approach: we allow the *temperature gradient* to enter the state equations, and we allow the temperature to affect the heat flux.[73] Precisely, writing

$$\mathbf{g} = \text{grad } T \qquad (14.13)$$

for the **temperature gradient**, we consider **bulk constitutive equations** in the form[74]

$$\varepsilon = \varepsilon(T, \mathbf{g}), \qquad \eta = \eta(T, \mathbf{g}), \qquad \mathbf{q} = \mathbf{q}(T, \mathbf{g}), \qquad (14.14)$$

---

[73] Such a starting assumption is based on Truesdell's principle of equipresence: 'a quantity present as an independent variable in one constitutive equation should be so present in all, unless... its presence contradicts some law of physics or rule of invariance' (cf. Truesdell and Noll (1965), Sect. 96). As Truesdell and Noll assert: 'This principle forbids us to eliminate any of the "causes" present from interacting with any other as regards a particular "effect". It reflects on the scale of gross phenomena the fact that all observed effects result from a common structure such as the motions of molecules.'

[74] For convenience, we assume that the material is *homogeneous*, so that the constitutive equations are independent of the material point. The equation $\varepsilon = \varepsilon(T, \mathbf{g})$ might, less ambiguously, be written in the form $\varepsilon = \hat{\varepsilon}(T, \mathbf{g})$ signifying that $\varepsilon(\mathbf{x}, t) = \hat{\varepsilon}(T(\mathbf{x}, t), \text{grad } T(\mathbf{x}, t))$ for all $\mathbf{x}$ and t.

which, by virtue of (14.10), yield a corresponding relation for the free energy:

$$\psi = \psi(T, \mathbf{g}) = \varepsilon(T, \mathbf{g}) - T\eta(T, \mathbf{g}). \tag{14.15}$$

We do not write a constitutive equation for the heat supply r, but instead allow r to be assignable in any way compatible with the basic laws, just as body forces are often left assignable in mechanics.

Given a smooth prescription $T(\mathbf{x}, t)$ of the temperature field over Ω for all time, the constitutive equations may be used to compute a corresponding *constitutive process* $(\varepsilon, \eta, \mathbf{q})$; the first law in the form (14.6) may then be used to determine the heat r that must be supplied to the body to support the process. The second law remains to be satisfied in all constitutive processes, a requirement that will result in restrictions on the constitutive equations.

We will refer to the constitutive equations as being **compatible with thermodynamics** if all constitutive processes are consistent with the second law in the form of the bulk free-energy inequality (14.11).

(14A) Compatibility theorem[75] *The constitutive equations are compatible with thermodynamics if and only if*:

(a) *the free energy, internal energy, and entropy are independent of $\mathbf{g}$ and satisfy*

$$\eta(T) = -\frac{\mathrm{d}\psi(T)}{\mathrm{d}T}, \qquad \varepsilon(T) = \psi(T) + T\eta(T); \tag{14.16}$$

(b) *the heat flux satisfies the heat-conduction inequality*

$$\mathbf{q}(T, \mathbf{g}) \cdot \mathbf{g} \leq 0. \tag{14.17}$$

PROOF. In view of the constitutive equations, (14.11) is equivalent to the inequality

$$[\psi_T(T, \mathbf{g}) - \eta(T, \mathbf{g})]T^{\cdot} + \psi_\mathbf{g}(\theta, \mathbf{g}) \cdot \mathbf{g}^{\cdot} + T^{-1}\mathbf{q}(T, \mathbf{g}) \cdot \mathbf{g} \leq 0; \tag{14.18}$$

thus the constitutive equations are compatible with thermodynamics if and only if (14.18) is satisfied by all time-dependent temperature fields $T = T(\mathbf{x}, t)$. The restrictions (a) and (b) imply (14.18). On the other hand, given any $\mathbf{x}_0$ and t_0, we can always find a temperature field T such that $T(\mathbf{x}_0, t_0)$, $\mathbf{g}(\mathbf{x}_0, t_0) = \operatorname{grad} T(\mathbf{x}_0, t_0)$, and $T^{\cdot}(\mathbf{x}_0, t_0)$ have arbitrarily assigned values. Thus the restrictions (a) and (b) are also necessary for compatibility with thermodynamics. ∎

We assume, for the remainder of the section, that the constitutive equations are compatible with thermodynamics.

---

[75] Coleman and Noll (1963); Coleman and Mizel (1963).

By (14.11) and (14.16), the entropy production is given by a function

$$\Gamma = \Gamma(T, \mathbf{g}) = -T^{-2}\mathbf{q}(T, \mathbf{g}) \cdot \mathbf{g} \geq 0, \qquad (14.19)$$

and this function has a minimum at $\mathbf{g} = \mathbf{0}$ for any choice of T. Thus the first derivative of $\Gamma(T, \mathbf{g})$ with respect to $\mathbf{g}$ vanishes at $\mathbf{g} = \mathbf{0}$, while the second derivative is positive semi-definite:

$$\mathbf{q}(T, \mathbf{0}) = \mathbf{0}, \qquad \mathbf{a} \cdot [\partial_{\mathbf{g}} \mathbf{q}(T, \mathbf{0})] \mathbf{a} \leq 0 \qquad (14.20)$$

for all vectors $\mathbf{a}$. The relation $(14.20)_1$ asserts the absence of a piezocaloric effect: *heat cannot flow in the absence of a temperature gradient.*

Consider now situations with temperature close to a given uniform temperature T_0 in the sense that

$$\delta = |T - T_0| + |\text{grad } T| \qquad (14.21)$$

is small. Let $\mathbf{K}$ denote the *conductivity tensor*

$$\mathbf{K} = -\partial_{\mathbf{g}} \mathbf{q}(T_0, \mathbf{0}) \qquad (14.22)$$

at T_0, so that, by (14.20), $\mathbf{K}$ is a positive semi-definite matrix. By (14.20), $\mathbf{q}_T(T_0, \mathbf{0}) = \mathbf{0}$; therefore, expanding $\mathbf{q}$ and Γ in Taylor series near $T = T_0$ and $\mathbf{g} = \mathbf{0}$, we arrive at the asymptotic relations

$$\mathbf{q} = -\mathbf{K} \text{ grad } T + O(\delta^2),$$
$$\Gamma = T_0^{-2} \text{ grad } T \cdot \mathbf{K} \text{ grad } T + O(\delta^3). \qquad (14.23)$$

For an *isotropic material* the tensor $\mathbf{K}$ is a scalar **conductivity** k times the identity matrix and $(14.23)_1$ reduces to

$$\mathbf{q} = -k \text{ grad } T + O(\delta^2), \qquad (14.24)$$

so that Fourier's law holds to within $O(\delta^2)$ near equilibrium.

The **bulk specific heat** is defined by

$$c(T) = \frac{d\varepsilon(T)}{dT}; \qquad (14.25)$$

in view of (14.16),

$$c(T) = T\frac{d\eta(T)}{dT} = -\frac{T\,d^2\psi(T)}{dT^2}, \qquad (14.26)$$

so that $c(T) > 0$ for all T if and only if the free energy is concave down.

Exercises

1. Determine the explicit form of the free energy and entropy for a material with constant specific heat.

2. Establish the Gibbs relations

$$\varepsilon^{\bullet} = T\eta^{\bullet}, \qquad \psi^{\bullet} = -\eta T^{\bullet}$$

and the entropy balance

$$T\eta^{\bullet} = -\operatorname{div} \mathbf{q} + r.$$

14.3 The heat equation

The constitutive equations (14.12) and the energy equation combine to form an important partial differential equation for the temperature:

$$c(T)T^{\bullet} = -\operatorname{div} \mathbf{q} + r, \qquad \mathbf{q} = \mathbf{q}(T, \operatorname{grad} T). \tag{14.27}$$

If we linearize this equation about the uniform state $T = T_0$, we find, using (14.27) and writing c for $c(T_0)$,

$$cT^{\bullet} = \operatorname{div}(\mathbf{K} \operatorname{grad} T) + r. \tag{14.28}$$

If the material is isotropic, then, because of (14.24), this linearization leads to the classical heat equation

$$cT^{\bullet} = k \Delta T + r, \tag{14.29}$$

with Δ the Laplacian in $\mathbb{R}^2$.

15
THERMODYNAMICS OF TWO-PHASE SYSTEMS[76]

We consider an **evolving interface** $\mathcal{S}$ separating two phases, labelled 1 and 2, and write $\Omega_1(t)$ and $\Omega_2(t)$ for the corresponding **phase regions** as discussed in Section 3.1. The normal $\mathbf{N}$ for $\mathcal{S}$ then coincides with the outward unit normal to $\partial\Omega_1$. For the moment we will restrict attention to situations in which the body $\Omega = \Omega_1(t) \cup \Omega_2(t)$ is all of $\mathbb{R}^2$.

We assume that the individual phases are immobile with bulk heat conduction the sole transport mechanism; we neglect conduction within the interface. We base most of the theory on balance of forces[77] in conjunction with the first two laws as stated in (14.1). To simplify the presentation we will omit from the discussion all supplies such as the heat supply r.

15.1 Basic quantities and the first two laws

The thermodynamics of the individual phases is described by the four **bulk fields** $T(\mathbf{x}, t)$, $\varepsilon(\mathbf{x}, t)$, $\eta(\mathbf{x}, t)$ and $\mathbf{q}(\mathbf{x}, t)$, defined in each of the bulk regions for all time. These fields are as discussed in Section 14.1, and, if we restrict attention to bulk control volumes R (cf. Chapter 3), then the first two laws in bulk are as stated in (14.3) and (14.5) (with $r = 0$) and lead to the relations (14.6)–(14.11) of Section 14.1 in the individual phase regions.

We assume that the bulk temperature

$$T \text{ is continuous across the interface}, \qquad (15.1)$$

and we consider T also as the *interfacial temperature*. We will not specify regularity hypotheses other than to note that:

$$\text{Bulk fields (other than temperature) are allowed to suffer jump discontinuities across the interface.} \qquad (15.2)$$

For φ any of the bulk fields, we write $[\varphi]$ for the **jump in** φ **across the interface**, phase 2 minus phase 1, as defined in (3.4).

We describe the thermodynamics of the interface in terms of three

---

[76] Cf. Gurtin (1988, 1991); Gurtin and Struthers (1990).
[77] As discussed in Chapter 4.

THERMODYNAMICS OF TWO-PHASE SYSTEMS   119

(scalar-valued) interfacial fields defined on the interface for all time:

$$e(\mathbf{x}, t) \quad \text{interfacial energy,}$$
$$s(\mathbf{x}, t) \quad \text{interfacial entropy,}$$
$$q(\mathbf{x}, t) \quad \text{apparent heat flux.}$$

An expression analogous to (14.10) then defines the **interfacial free energy**

$$f = e - Ts. \tag{15.3}$$

Let R be a two-phase control volume (cf. Chapter 3), with $\imath(t)$ the portion of the interface $\jmath(t)$ that lies in R. Then

$$\int_R \varepsilon \, da + \int_\imath e \, ds, \quad \int_R \eta \, da + \int_\imath s \, ds$$

give the internal energy and the internal entropy of R, while (14.2) and (14.4)$_1$ represent the net heat and entropy flow into R by bulk conduction across ∂R.

Unlike the single-phase theory discussed in the last chapter, here there is power expended by the motion of the interface. As in the mechanical theory, we assume that this power expense is represented by

$$\int_{\partial \imath} \mathbf{C} \cdot \mathbf{v}_{\partial \imath} = \int_{\partial \imath} \sigma v_{\partial \imath (\tan)} + \int_{\partial \imath} \xi V$$

(cf. (4.6) and the discussion given in the paragraph following (5.2)). One manifestation of surface tension is through the power expense

$$\int_{\partial \imath} \sigma v_{\partial \imath (\tan)}$$

as material is added to the edge of the evolving subcurve $\imath$. In this spirit, we assume that this addition of material results in apparent heat and entropy flows

$$\int_{\partial \imath} q v_{\partial \imath (\tan)}, \quad \int_{\partial \imath} \left(\frac{q}{T}\right) v_{\partial \imath (\tan)}. \tag{15.4}$$

The first two laws[78] for R are then **balance of energy**

$$\frac{d}{dt}\left\{\int_R \varepsilon \, da + \int_\imath e \, ds\right\} = -\int_{\partial R} \mathbf{q} \cdot \mathbf{n} \, ds + \int_{\partial \imath} q v_{\partial \imath (\tan)} + \int_{\partial \imath} \mathbf{C} \cdot \mathbf{v}_{\partial \imath} \tag{15.5}$$

[78] Gurtin (1988, 1991); Gurtin and Struthers (1990). The earliest attempt to state the first two laws for a moving phase interface is apparently that of Moeckel (1975), but as noted by Fernandez-Diaz and Williams (1979), Moeckel's work has serious errors. Fernandez-Diaz and Williams essentially formulate (15.5) (without the integrals over $\partial \imath$), but, as noted by Williams (private communication, 1985), their outflow term, which replaces the apparent heat supply, is incorrect. Other discussions are given by Wollkind (1979); Rogers (1983); Caroli et al. (1984); Gurtin (1986).

THERMODYNAMICS OF TWO-PHASE SYSTEMS

and **growth of entropy**

$$\frac{d}{dt}\left\{\int_R \eta\, da + \int_\imath s\, ds\right\} \geq -\int_{\partial R}\left(\frac{\mathbf{q}}{T}\right)\cdot \mathbf{n}\, ds + \int_{\partial \imath}\left(\frac{q}{T}\right)v_{\partial \imath(\tan)}; \quad (15.6)$$

these laws are required to hold for all two-phase control volumes R.

If the process in question is *isothermal* ($T \equiv$ constant), then, multiplying (15.6) by T and subtracting the resulting equation from (15.5), we are led, using (14.10) and (15.3), to the inequality

$$\frac{d}{dt}\left\{\int_R \psi\, da + \int_\imath f\, ds\right\} \leq \int_{\partial \imath} \mathbf{C}\cdot \mathbf{v}_{\partial \imath}, \quad (15.7)$$

which is the dissipation inequality (5.3) upon which the mechanical theory discussed in Part II was based.

(15A) Remark Assume that the control volume R contains the interface, so that $\imath = \mathfrak{s}$, $\partial\imath = \varnothing$. If R is *insulated* ($\mathbf{q}\cdot\mathbf{n} = 0$ on ∂R), then (15.5) and (15.6) imply

$$\begin{aligned}\frac{d}{dt}\left\{\int_R \varepsilon\, da + \int_\mathfrak{s} e\, ds\right\} &= 0, \\ \frac{d}{dt}\left\{\int_R \eta\, da + \int_\mathfrak{s} s\, ds\right\} &\geq 0.\end{aligned} \quad (15.8)$$

Thus for R insulated the internal energy of R is constant and the entropy of R is non-decreasing. If, on the other hand, ∂R is *isothermal* ($T = T_0 =$ constant on ∂R), then (15.5) and (15.6) may be combined to yield

$$\frac{d}{dt}\left\{\int_R W\, da + \int_\mathfrak{s} w\, ds\right\} \leq 0, \quad (15.9)$$

with W and w the bulk and interfacial Gibbs functions

$$W = \varepsilon - T_0\eta, \qquad w = e - T_0 s. \quad (15.10)$$

These observations have implications concerning the analysis of corresponding free-boundary problems, since they suggest the existence of Lyapunov functions that decrease with time on solution paths.

Exercises

1. Derive (15.7) and (15.9).

2. Let $u = (T - T_0)/T$. Show that, granted (15.5), (15.6) is equivalent to the

inequality[79]

$$\frac{d}{dt}\left\{\int_R W\, da + \int_\imath w\, ds\right\} \leq -\int_{\partial R} u\mathbf{q}\cdot\mathbf{n}\, ds + \int_{\partial \imath} u\mathscr{q} v_{\partial \imath(\tan)} + \int_{\partial \imath} \mathbf{C}\cdot\mathbf{v}_{\partial \imath}.$$

(15.11)

Returning to the general theory, choose an evolving subcurve $\imath$ of $\mathscr{s}$ and an arbitrary time t_0, apply (15.5) and (15.6) to a pillbox $R(\varepsilon)$, $0 < \varepsilon < \varepsilon_0$, that shrinks to $\imath(t_0)$, let $\varepsilon \to 0$, and use (3.6) and the fact that t_0 is arbitrary; the results are the **interfacial laws**:

$$\frac{d}{dt}\int_\imath e\, ds = \int_\imath [\varepsilon] V\, ds - \int_\imath [\mathbf{q}]\cdot\mathbf{N}\, ds + \int_{\partial \imath} \mathscr{q} v_{\partial \imath(\tan)} + \int_{\partial \imath} \mathbf{C}\cdot\mathbf{v}_{\partial \imath},$$

$$\frac{d}{dt}\int_\imath s\, ds \geq \int_\imath [\eta] V\, da - \int_\imath \left[\frac{\mathbf{q}}{T}\right]\cdot\mathbf{N}\, ds + \int_{\partial \imath} \left(\frac{\mathscr{q}}{T}\right) v_{\partial \imath(\tan)}.$$

(15.12)

These laws depend on the choice of the control volume only through the evolving subcurve $\imath$ of $\mathscr{s}$; in fact, *they are valid for all evolving subcurves.*

Granted the laws (14.3) and (14.5) in bulk, the interfacial laws (15.12) are equivalent to (15.5) and (15.6). The interfacial laws are most easily interpreted by considering $\imath$ as an infinitesimally thin evolving two-phase control volume. Then

$$\int_\imath e\, ds, \qquad \int_\imath s\, ds$$

(15.13)

represent the energy and entropy of $\imath(t)$;

$$\int_\imath [\varepsilon] V\, ds, \qquad \int_\imath [\eta] V\, ds$$

(15.14)

represent flows of bulk energy and bulk entropy into $\imath(t)$;

$$-\int_\imath [\mathbf{q}]\cdot\mathbf{N}\, da, \qquad -\int_\imath \left[\frac{\mathbf{q}}{T}\right]\cdot\mathbf{N}\, da$$

represent heat and entropy transferred to $\imath(t)$ from the bulk material by heat conduction;

$$\int_{\partial \imath} \mathscr{q} v_{\partial \imath(\tan)}, \qquad \int_{\partial \imath} \left(\frac{\mathscr{q}}{T}\right) v_{\partial \imath(\tan)}$$

(15.15)

account for heat and entropy flow into $\imath(t)$ induced by the tangential

---

[79] Cf. Gurtin (1992). The relations (15.5) and (15.11) formally represent balance of mass and a free-energy inequality for isothermal mass transport in a two-phase system with ε and e the bulk and interfacial concentrations, **q** the mass flux, $\mathscr{q}$ the apparent mass flux, u the chemical potential, W and w the bulk and interfacial free energies, and **C** the capillary force (cf. Gurtin (1991)).

motion of $\partial\imath(t)$; and

$$\int_{\partial\imath} \mathbf{C}\cdot\mathbf{v}_{\partial\imath} \tag{15.16}$$

gives the power expended on $\imath(t)$ by the capillary force acting across $\partial\imath(t)$.

15.2 Local forms of the interfacial laws

If we apply the transport theorem (2.24) to (15.12) and use (4.4)$_3$ and the power identity (4.9), we find that

$$\int_{\imath} \{e^\circ - \xi\theta^\circ - (e - \sigma)KV + (B - [\varepsilon])V + [\mathbf{q}]\cdot\mathbf{N}\}\, ds$$

$$+ \int_{\partial\imath} \{e - \mathscr{g} - \sigma\}v_{\partial\imath(\text{tan})} = 0,$$

(15.17)

$$\int_{\imath} \{s^\circ - sKV - [\eta]V + T^{-1}[\mathbf{q}]\cdot\mathbf{N}\}\, ds + \int_{\partial\imath} \{(s - (T^{-1}\mathscr{g})\}v_{\partial\imath(\text{tan})} \geq 0,$$

for all evolving subcurves $\imath$ of $\mathscr{s}$, where B is the normal interaction (5.9). These relations and (3B) yield the **edge balances**

$$e = \sigma + \mathscr{g}, \qquad s = \frac{\mathscr{g}}{T}, \tag{15.18}$$

which express balance of energy and entropy associated with the addition of material at the edge of an evolving subcurve. An important consequence of the edge balances is the next theorem.

(15B) Tension–energy theorem *The surface tension and the interfacial free energy coincide*:

$$\sigma = f. \tag{15.19}$$

If we again apply (3B) to (15.17)$_1$, and use (15.18), we are led to

$$e^\circ - \xi\theta^\circ - TsKV + (B - [\varepsilon])V + [\mathbf{q}]\cdot\mathbf{N} = 0, \tag{15.20}$$

or equivalently, by (2.9)$_2$, (4.4)$_1$, (15.3), and (15.19), to

$$[\varepsilon]V = [\mathbf{q}]\cdot\mathbf{N} + e^\circ - eKV - (\xi V)_s, \tag{15.21}$$

which we will refer to as the **local energy balance**. Similarly, (15.17)$_2$ yields the **local entropy-growth inequality**

$$s^\circ - sKV - [\eta]V + T^{-1}[\mathbf{q}]\cdot\mathbf{N} \geq 0, \tag{15.22}$$

and subtracting (15.22) multiplied by T from (15.20), we arrive with the aid of (14.10) and (15.3), at the **interfacial free-energy inequality**:

$$f^\circ - \xi\theta^\circ + sT^\circ + (B - [\psi])V \leq 0. \tag{15.23}$$

This inequality will play a role analogous to that played in the mechanical theory by (5.8) in determining appropriate constitutive equations for the interface.

The interfacial field

$$\gamma = s^\circ - sKV - [\eta]V + T^{-1}[\mathbf{q}]\cdot\mathbf{N} \geq 0 \tag{15.24}$$

represents the **interfacial entropy production**. Using γ and the bulk entropy production Γ as defined in each phase by (14.8), we can write the difference between the left and right sides of the entropy inequality (15.6) as

$$\int_R \Gamma\, da + \int_\imath \gamma\, ds. \tag{15.25}$$

In addition,

$$-T\gamma = f^\circ - \xi\theta^\circ + sT^\circ + (B - [\psi])V. \tag{15.26}$$

Exercise Establish (15.25) and (15.26).

16
CONSTITUTIVE THEORY[80]

16.1 Constitutive equations for the bulk material

We consider, for each of the two phases ($i = 1, 2$), **bulk constitutive equations**

$$\varepsilon = \varepsilon_i(T, \mathbf{g}), \qquad \eta = \eta_i(T, \mathbf{g}), \qquad \mathbf{q} = \mathbf{q}_i(T, \mathbf{g}), \qquad (16.1)$$

with $\mathbf{g} = \text{grad } T$ the **temperature gradient**. These yield an analogous relation for the bulk free energy:

$$\psi = \psi_i(T) = \varepsilon_i(T) - T\eta_i(T). \qquad (16.2)$$

The determination of constitutive restrictions necessary and sufficient to ensure that the bulk free-energy inequality (14.11) hold in all 'constitutive processes' is as in Section 14.2: the results are the requirements that $\psi_i(T, \mathbf{g})$, $\varepsilon_i(T, \mathbf{g})$, and $\eta_i(T, \mathbf{g})$ be independent of $\mathbf{g}$, and that

$$\eta_i(T) = -\frac{\mathrm{d}\psi_i(T)}{\mathrm{d}T}, \qquad \mathbf{q}_i(T, \mathbf{g}) \cdot \mathbf{g} \leq 0. \qquad (16.3)$$

The results of Sections 14.2 and 14.3 have analogues within the two-phase theory. In particular, when the temperature is close to a given uniform temperature T_0, in the sense that δ defined by (14.21) is small, the heat flux has the asymptotic form

$$\mathbf{q} = -\mathbf{K}_i \text{ grad } T + O(\delta^2), \qquad (16.4)$$

with $\mathbf{K}_i$, the **conductivity tensor** for phase i at the temperature T_0, a positive semi-definite 2×2 matrix. Under isotropy this reduces to

$$\mathbf{q} = -k_i \text{ grad } T + O(\delta^2), \qquad (16.5)$$

with k_i the scalar **conductivity** at T_0. Further, if we define the **bulk specific heats** by

$$c_i(T) = \frac{\mathrm{d}\varepsilon_i(T)}{\mathrm{d}T}, \qquad (16.6)$$

then we are led to heat equations of the form (14.27)–(14.29) to be satisfied in each phase.

[80] Cf. Gurtin (1988).

16.2 The transition temperature

In classical theories of melting and freezing there is a temperature T_M, called the melting temperature, at which the phase transition takes place. At T_M the free energies of the two phases coincide, with the solid having lower free energy for $T < T_M$, the liquid having lower free energy for $T > T_M$. Because of this, the temperature is generally used to characterize the individual phases, the solid being restricted to temperatures $T < T_M$, the liquid to temperatures $T > T_M$. Here we suppose that the bulk free energies of the individual phases coincide at a single temperature T_M, called the transition temperature, but *we do not require that the phase change take place at* T_M; instead we allow each of the phases to have temperatures above and below T_M, so that *supercooling of the liquid* ($T < T_M$) and *superheating of the solid* ($T > T_M$) are possible.

Precisely, we assume that there is a *unique* temperature T_M, called the **transition temperature**, at which the bulk free energies coincide:

$$\psi_1(T_M) = \psi_2(T_M). \tag{16.7}$$

The difference

$$\ell = \varepsilon_2(T_M) - \varepsilon_1(T_M) \tag{16.8}$$

in energy between phases at the transition temperature is the **latent heat**, which we assume to be non-zero:

$$\ell \neq 0. \tag{16.9}$$

By (16.2), (16.7), and (16.8),

$$\ell = T_M\{\eta_2(T_M) - \eta_1(T_M)\}. \tag{16.10}$$

To discuss behaviour near the transition temperature, we introduce the (dimensionless) **temperature difference**

$$u = \frac{T - T_M}{T_M}. \tag{16.11}$$

Then, by (16.3), (16.7), (16.8), and (16.10), for u small,

$$\psi_2(T) - \psi_1(T) = -\ell u + O(u^2),$$
$$\varepsilon_2(T) - \varepsilon_1(T) = \ell + O(u), \tag{16.12}$$
$$\eta_2(T) - \eta_1(T) = \frac{\ell}{T_M} + O(u).$$

(16A) Remark Assume that $\ell > 0$, which would be the case were phase 1 solid and phase 2 liquid. Then, by (16.12)$_1$,

$$\psi_1(T) > \psi_2(T) \quad \text{if } T > T_M,$$
$$\psi_1(T) < \psi_2(T) \quad \text{if } T < T_M, \tag{16.13}$$

so that *the liquid is the more stable phase at temperatures above the transition temperature, while the solid is more stable at temperatures below the transition temperature.*

16.3 Constitutive equations for the interface

As **interfacial constitutive equations** we allow the free energy, the entropy, the capillary force, and the normal interaction to depend on the thermal behaviour through a dependence on T, and on the orientation and kinetics of the interface through a dependence on θ and V:

$$f = f(T, \theta, V), \quad s - s(T, \theta, V), \quad \mathbf{C} = \mathbf{C}(T, \theta, V), \quad B = B(T, \theta, V). \tag{16.14}$$

By (1.3), (4.1), (4.4)$_3$, and (15.3), these relations induce additional constitutive equations

$$\sigma = \sigma(T, \theta, V), \quad \xi = \xi(T, \theta, V), \quad m = m(T, \theta, V), \quad e = e(T, \theta, V); \tag{16.15}$$

in fact, (4.4)$_3$ and (15.19) imply that

$$\sigma(T, \theta, V) = f(T, \theta, V), \quad m(T, \theta, V) = -\xi(T, \theta, V). \tag{16.16}$$

Given an evolving interface $\mathscr{S}$ and a time-dependent temperature field T, the constitutive equations may be used to compute a corresponding **constitutive process** $(f, s, \mathbf{C}, B)$. Without suitable restrictions on the constitutive equations such processes will not all satisfy the interfacial free-energy inequality (15.23). When all constitutive processes are consistent with (15.23) we will refer to the constitutive equations as **compatible with thermodynamics**.

(16B) Compatibility theorem *The constitutive equations are compatible with thermodynamics if and only if:*

(a) *the free energy, entropy, surface tension, and surface shear are independent of V and satisfy*

$$\sigma(T, \theta) = f(T, \theta), \quad \xi(T, \theta) = f_\theta(T, \theta), \quad s(T, \theta) = -f_T(T, \theta); \tag{16.17}$$

(b) *the normal interaction has the form*

$$B(T, \theta, V) = \psi_2(T) - \psi_1(T) - b(T, \theta, V)V,$$
$$b(T, \theta, V) \geq 0. \tag{16.18}$$

Exercise Prove the compatibility theorem (cf. the proof of (6A)).

(16C) Remark By (1.3) and (16.17), the *capillary force* (4.1) has the form

$$\mathbf{C} = \mathbf{C}(T, \theta) = f(T, \theta)\mathbf{T}(\theta) + f_\theta(T, \theta)\mathbf{N}(\theta). \tag{16.19}$$

The relations (1.5), (4.4)$_2$, and (16.17) imply that, in every constitutive process,

$$B_{\text{tan}} = \mathsf{s}(T, \theta)T_s, \tag{16.20}$$

so that variations in temperature along the interface require the presence of *tangential interactive forces*. This is in contrast to the purely mechanical theory in which tangential forces are not needed (cf. the remark following (6.10)).

Note that (15.3) and (16.17) yield the **Gibbs relations**

$$f^\circ = -\mathsf{s}T^\circ + \xi\theta^\circ, \qquad e^\circ = T\mathsf{s}^\circ + \xi\theta^\circ, \tag{16.21}$$

Note also that, by (16.18) and (16.21)$_2$, we can write the energy balance (15.20) as an entropy balance

$$T[\eta]V = [\mathbf{q}] \cdot \mathbf{N} + T\mathsf{s}^\circ - T\mathsf{s}KV - bV^2. \tag{16.22}$$

This relation, (14.19), and (15.24) imply that the bulk and interfacial entropy productions satisfy

$$\Gamma = -T^{-2}\mathbf{q} \cdot \mathbf{g}, \qquad \gamma = T^{-1}bV^2. \tag{16.23}$$

17

FREE-BOUNDARY PROBLEMS[81]

17.1 Bulk equations and interface conditions

The equations derived thus far combine to form an important free-boundary problem for the temperature. The differential equation to be satisfied in bulk is balance of energy supplemented by the constitutive equations (cf. (14.27)):

$$c_i(T)T^{\bullet} = -\text{div}\,\mathbf{q}, \qquad \mathbf{q} = \mathbf{q}_i(T, \text{grad}\,T). \qquad (17.1)$$

The evolving interface forms a free boundary, and the corresponding free-boundary conditions follow from the force and energy balances for the interface. Energy balance is given by (15.20), while (4.4)$_1$, (15.19), (16.17), and (16.18) yield the following expression for the normal force balance:

$$[\psi] = -fK - \xi_s + bV, \qquad (17.2)[82]$$

with $b = b(T, \theta)$ assumed independent of V.

The basic equations which govern the evolution of the interface consist of the bulk equations (17.1), the interface conditions (15.20) and (17.2), and the appropriate constitutive equations:

bulk equations

$$c_i(T)T^{\bullet} = -\text{div}\,\mathbf{q}, \qquad \mathbf{q} = \mathbf{q}_i(T, \text{grad}\,T),$$

$$\psi = \psi_i(T), \qquad \varepsilon = \varepsilon_i(T) = \psi_i(T) - T\frac{\mathrm{d}\psi_i(T)}{\mathrm{d}T}, \qquad (17.3)$$

interface conditionsy

$$[\psi] = -fK - \xi_s + bV,$$

$$[\varepsilon]V = [\mathbf{q}]\cdot\mathbf{N} + e^{\circ} - eKV - (\xi V)_s,$$

$$f = f(T, \theta), \qquad e = e(T, \theta) = f(T, \theta) - Tf_T(T, \theta), \qquad (17.4)$$

$$\xi = \xi(T, \theta) = f_{\theta}(T, \theta), \qquad b = b(T, \theta).$$

(17A) Remark *In the absence of interfacial structure* (that is, for f, ξ, and b identically zero) (17.4)$_1$ yields $[\psi] = 0$, so that, by (16.7),

$$T = T_M. \qquad (17.5)$$

---

[81] Cf. Gurtin (1986, 1988).
[82] Within a *statical theory* ($V = 0$) Herring (1951a) and Cahn and Hoffman (1974) derive an equation of this form as a necessary condition for the free energy to be a minimum. The assumption that b be independent of V is made for convenience only.

This is a free-boundary condition of the classical (Stefan) theory of melting. In that theory the solid is required to have temperatures $T < T_M$, the liquid temperatures $T > T_M$; within the more general theory developed here (17.5) remains valid, even when the solid is superheated and the liquid supercooled, provided the interface is unstructured. Further, when interfacial structure is included, we may use (16.17), (17.2), and the constitutive equations to conclude that an isothermal, flat, and stationary interface has $T = T_M$, but a non-isothermal, curved, and moving interface need not.

17.2 Initial conditions and boundary conditions

Appropriate **initial conditions** involve the prescription of

$$\Omega_1(0) \quad \text{and} \quad T(\mathbf{x}, 0) \quad \text{for all } \mathbf{x} \in \mathbb{R}^2. \tag{17.6}$$

Since the body (the region of space occupied by the two phases) is all of $\mathbb{R}^2$, conditions at infinity are needed; these are standard if the interface is finite.

Thus far we have limited our discussion to unbounded bodies. If the **body** Ω is a *bounded region* (fixed in time), then boundary conditions are required. When the interface $\mathcal{A}(t)$ touches the boundary, conditions expressing balance of capillary forces are needed at the juncture of the interface and the boundary; these require a detailed description of the *interface* between the individual phases and $\partial\Omega$. Here we will restrict attention to situations in which the interface does not touch the boundary; in the same spirit, when discussing boundary conditions away from $\mathcal{A}(t)$, we will ignore the effects of the interface defined by $\partial\Omega$. Appropriate boundary conditions are then a prescription of

$$\begin{aligned}
& T(\mathbf{x}, t) && \text{on a portion of } \partial\Omega \text{ and} \\
& \mathbf{q}(\mathbf{x}, t) \cdot \mathbf{n}(\mathbf{x}) && \text{on the remainder, with} \\
& \mathbf{n} && \text{the outward unit normal to } \partial\Omega.
\end{aligned} \tag{17.7}$$

The free-boundary problem described by (17.3), (17.4), (17.6), and (17.7) is extremely difficult, chiefly because of the non-linearities inherent in the free-boundary conditions (17.4). For that reason we will develop, in the next section, an approximate theory for weak interfaces.

(17B) Remark Note that (15.8) and (15.9) with $R = \Omega$ yield Lyapunov functions and associated global growth relations for a body whose boundary is either insulated or isothermal.

17.3 Free-boundary problems near the transition temperature for weak interfaces

We now derive approximate free-boundary problems under the assumption that the interfacial free energy and kinetic coefficient are small, and that the temperature is close to the transition temperature. The derivation we give is purely formal.

Let δ denote a small dimensionless quantity. We assume that:[83]

$f(T, \theta)$, $b(T, \theta)$, and hence $e(T, \theta)$, $s(T, \theta)$, $\xi(T, \theta)$, are strongly $O(\delta)$;

$\psi_i(T)$, $\mathbf{q}_i(T, \text{grad } T)$, and hence $\varepsilon_i(T)$, $\eta_i(T)$, are strongly $O(1)$;

and we will consider solutions of (17.3) and (17.4) that have:

$u(\mathbf{x}, t)$ strongly $O(\delta)$;

$K(\mathbf{x}, t)$, $V(\mathbf{x}, t)$ strongly $O(1)$,

where u is the temperature difference (16.11).

17.3.1 Approximate interface conditions

Let

$$f_0(\theta) = f(T_M, \theta), \qquad \xi_0(\theta) = \xi(T_M, \theta), \qquad b_0(\theta) = b(T_M, \theta). \quad (17.8)$$

Then, in view of (16.12), the interface conditions (17.4) have the asymptotic forms

$$\begin{aligned} \ell u &= f_0(\theta)K + \xi_0(\theta)_s - b_0(\theta)V + O(\delta^2), \\ \ell V &= [\mathbf{q}] \cdot \mathbf{N} + O(\delta). \end{aligned} \quad (17.9)$$

By $(16.17)_2$,

$$\xi_0(\theta) = f'_0(\theta);$$

thus neglecting higher-order terms in (17.9), and dropping the subscript zero, we are led to the **approximate interface conditions**

$$\begin{aligned} \ell u &= \{f(\theta) + f''(\theta)\}K - b(\theta)V, \\ \ell V &= [\mathbf{q}] \cdot \mathbf{N}, \end{aligned} \quad (17.10)[84]$$

where we have used (1.5).

---

[83] $g(z)$ is strongly $O(\delta)$ if $g(z)$ and all of its derivatives are uniformly $O(\delta)$. We assume that the equations (17.3) and (17.4) have been scaled to render the underlying quantities dimensionless.
[84] Cf. Gurtin (1988), eqn (6.9). The boundary condition $(17.10)_2$ is a classical Stefan condition. Free-boundary conditions of the form $\ell u = -b(\theta)V$ were introduced by Frank (1958) and used by Chernov (1963a,b); $\ell u = fK$ was introuduced by Mullins (1960) and Mullins and Sekerka (1963, 1964); $\ell u = -bV + fK$ was use by Voronkov (1964). See also Seidensticker (1966), Tarshis and Tiller (1967), Gurtin (1986), and Visintin (1988, 1989), and the review articles by Sekerka (1968, 1973, 1984), Chernov (1971, 1974), Delves (1974), and Langer (1980).

17.3.2 Approximate free-boundary problems

To within terms of $O(\delta)$, (17.3) are approximated by the linear equations:

$$c_i T^{\bullet} = -\operatorname{div} \mathbf{q}, \qquad \mathbf{q} = -\mathbf{K}_i \operatorname{grad} T,$$

with $c_i = c_i(T_M)$ the bulk specific heat (16.6) for phase i at the transition temperature, and $\mathbf{K}_i$ the conductivity tensor for phase i at the transition temperature (cf. (16.4)). The system we will consider consists of these equations in bulk together with the interface conditions (17.10).

We label phases so that *phase 2 has the higher internal energy at the transition temperature*. Then[85]

$$\ell > 0, \qquad (17.11)$$

and to avoid an unnecessary constant, we define $c_i^* = c_i T_M/\ell$, $\mathbf{q}^* = \mathbf{q}/\ell$, $\mathbf{K}_i^* = \mathbf{K}_i T_M/\ell$, $f^* = f/\ell$, and $b^* = b/\ell$, and then drop the star superscript. In addition, we write[86]

$$g(\theta) = f(\theta) + f''(\theta). \qquad (17.12)$$

We are then led to the **quasi-linear equations**:[87]

$$\begin{aligned}
c_i u^{\bullet} &= -\operatorname{div} \mathbf{q}, & \mathbf{q} &= -\mathbf{K}_i \operatorname{grad} u & &\text{in bulk}, \\
u &= g(\theta) K - b(\theta) V, & V &= [\mathbf{q}] \cdot \mathbf{N} & &\text{on the interface},
\end{aligned} \qquad (17.13)$$

where, repeating assumptions made earlier,

$$b(\theta) \geq 0, \qquad \mathbf{K}_i \text{ is positive semi-definite}.$$

The **quasi-linear problem**[88] consists of (17.13) supplemented by the initial conditions (17.6) and the boundary conditions (17.7) (with T replaced by u and $\mathbf{q} = -\mathbf{K}_i \operatorname{grad} u$).

Generally, one expects the interface to move slowly in comparison to the time scale for heat conduction. With this in mind, we consider the **quasi-static equations** in which the terms $c_i u^{\bullet}$ in the bulk equations are neglected:

$$\begin{aligned}
\operatorname{div} \mathbf{q} &= 0, & \mathbf{q} &= -\mathbf{K}_i \operatorname{grad} u & &\text{in bulk}, \\
u &= g(\theta) K - b(\theta) V, & V &= [\mathbf{q}] \cdot \mathbf{N} & &\text{on the interface}.
\end{aligned} \qquad (17.14)$$

---

[85] So that for a solid–liquid system phase 1 represents the solid, phase 2 the liquid (cf. Remark 16A).
[86] Note that this definition of $g(\theta)$ differs from (8.6).
[87] In bulk means in $\Omega_i(t)$ for all t and $i = 1, 2$; on the interface means on $\mathscr{I}(t)$ for all t.
[88] Chen and Reitich (1992) establish local existence and uniqueness for the isotropic problem with $f, b > 0$. The corresponding problem for $b = 0$ is open, although Luckhaus (1990) has established a global existence theorem for 'weak' solutions, and a similar result has been announced by Almgren and Wang (private communication).

The **quasi-static problem**[89] consists of (17.14) supplemented by $(17.6)_1$ and (17.7). (The condition $(17.6)_2$, involving the prescription of $u(\mathbf{x}, 0)$, is dropped.) If the body is infinite, the boundary conditions (17.7) are replaced by conditions at infinity.

In discussing the above problems, it is tacit that the interface does not touch $\partial\Omega$; in particular, the initial data must be consistent with this assumption.

17.3.3 The first two laws for the approximate theories

The quasi-linear and quasi-static equations were obtained as approximations of the general theory and therefore cannot be expected to obey the general laws of energy balance and entropy growth. Interestingly, these theories are accompanied by their own versions of the thermodynamical laws. To state these precisely, we assume that the specific heats of the two phases coincide ($c_1 = c_2 = c$), and—for any solution of the quasi-linear equations—we define an *internal energy*

$$\varepsilon(\mathbf{x}, t) = \begin{cases} cu(\mathbf{x}, t), & \mathbf{x} \in \Omega_1(t) \\ 1 + cu(\mathbf{x}, t), & \mathbf{x} \in \Omega_2(t) \end{cases} \tag{17.15}$$

and a *Gibbs function*

$$\varphi(\mathbf{x}, t) = \tfrac{1}{2}cu(\mathbf{x}, t)^2. \tag{17.16}$$

In addition, we now write $\mathbf{C}(\theta)$ for the *capillary force at the transition temperature*, so that

$$\mathbf{C}(\theta) = f(\theta)\mathbf{T}(\theta) + f'(\theta)\mathbf{N}(\theta) \tag{17.17}$$

(cf. (6.10)), and, for R a two-phase control volume, we denote by $\imath(t)$ and $R_i(t)$, respectively, the portions of the interface and the phase region $\Omega_i(t)$ that lie in R (cf. (3.2), (3.3)).

(17C) Thermodynamical laws for the quasi-linear equations[90] *Consider a solution of the quasi-linear equations with $c_1 = c_2 = c$. Then for any two-phase control volume R,*

$$\frac{d}{dt}\int_R \varepsilon\, da = -\int_{\partial R} \mathbf{q}\cdot\mathbf{n}\, ds,$$

$$\frac{d}{dt}\left\{\int_R \varphi\, da + \int_\imath f(\theta)\, da\right\} \leq -\int_{\partial R} u\mathbf{q}\cdot\mathbf{n}\, ds + \int_{\partial\imath} \mathbf{C}\cdot v_{\partial\imath}, \tag{17.18}$$

---

[89] Brush and Sekerka (1989) discuss numerical solutions for isotropic bulk behaviour, but an anisotropic interface. Duchon and Robert (1984) establish local existence and uniqueness for the isotropic problem with $b = 0$ and vanishing conductivity in one of the two phases.

[90] Cf. Gurtin (1992), who derives 17C when interfacial structure is neglected and gives arguments asserting that for $c_1 \neq c_2$ the quasi-linear problem could give erroneous results.

the difference between the right and left sides of $(17.18)_2$ being

$$\int_\imath b(\theta)V^2 \, ds + \sum_{i=1,2} \int_{R_i} \text{grad } u \cdot \mathbf{K}_i \, \text{grad } u \, da \qquad (17.19)$$

PROOF. The proof is based on four identities. The first, a direct consequence of the divergence theorem, asserts that

$$\sum_{i=1,2} \int_{R_i} \text{div } \mathbf{h} \, da = \int_{\partial R} \mathbf{h} \cdot \mathbf{n} \, ds - \int_\imath [\mathbf{h}] \cdot \mathbf{N} \, ds \qquad (17.20)$$

for any bulk vector field $\mathbf{h}$. The second,

$$\left\{ \int_R u^p \, ds \right\}^\cdot = \int_R (u^p)^\cdot \, ds, \qquad (17.21)$$

$p = 1, 2$, follows from the continuity of u across the interface. The final two identities are:

$$\int_\imath V \, ds = \text{area}(R_1)^\cdot = -\text{area}(R_2)^\cdot,$$

$$\frac{d}{dt} \int_\imath f(\theta) \, da = -\int_\imath g(\theta) K V \, ds + \int_{\partial \imath} \mathbf{C}(\theta) \cdot \mathbf{v}_{\partial \imath}. \qquad (17.22)$$

The first of (17.22) follows from $(2.26)_1$. In view of $(2.9)_2$ and (2.24),

$$\frac{d}{dt} \int_\imath f(\theta) \, da = \int_\imath [f'(\theta)V_s - f(\theta)KV] \, ds + \int_{\partial \imath} f(\theta)\mathbf{v}_{\partial \imath(\text{tan})},$$

and integrating the term involving $f'(\theta)V_s$ by parts we conclude, with the aid of (1.5), (2.11), $(2.16)_1$, (17.12), and (17.17), that $(17.22)_2$ is valid.

To establish the energy balance $(17.18)_1$, we use $(17.13)_{1,4}$, (17.15), (17.21), and $(17.22)_1$ to show that

$$\frac{d}{dt}\int_R \varepsilon \, da = \int_R cu^\cdot \, ds - \int_\imath V \, ds = -\int_\imath [\mathbf{q}] \cdot \mathbf{N} \, ds - \sum_{i=1,2} \int_{R_i} \text{div } \mathbf{q} \, da; \qquad (17.23)$$

this and (17.20) yield $(17.18)_1$.

To verify $(17.18)_2$, note first that, by $(17.13)_1$ and (17.16),

$$\varphi^\cdot = -\text{div}(u\mathbf{q}) + \mathbf{q} \cdot \text{grad } u \quad \text{in bulk,}$$

and this relation with (17.22) yields

$$\frac{d}{dt}\int_R \varphi \, da = -\int_{\partial R} u\mathbf{q} \cdot \mathbf{n} \, ds + \int_\imath u[\mathbf{q}] \cdot \mathbf{N} \, ds + \int_R \mathbf{q} \cdot \text{grad } u \, da.$$

On the other hand, by $(17.13)_{3,4}$ and $(17.22)_2$,

$$\int_{\imath} u[\mathbf{q}]\cdot \mathbf{N}\,ds = -\frac{d}{dt}\int_{\imath} f(\theta)\,da + \int_{\partial \imath} \mathbf{C}(\theta)\cdot \mathbf{v}_{\partial \imath} - \int_{\imath} b(\theta)V^2\,ds.$$

The last two relations and $(17.13)_2$ imply $(17.18)_2$. ∎

Exercise Show that, to within the approximation of a weak interface, the expression (17.19) is proportional to the total entropy production in R.

Note that, by $(17.22)_1$ and (17.23),

$$\frac{d}{dt}\int_R \varepsilon\,da = \int_R cu^{\cdot}\,ds - \text{area}(R_1)^{\cdot}. \tag{17.24}$$

The quasi-static equations are formally equivalent to the quasi-linear equations with $c_1 = c_2 = 0$; thus the quasi-static equations correspond to $\varphi(\mathbf{x}, t) \equiv 0$ and to the internal energy

$$\varepsilon(\mathbf{x}, t) = \begin{cases} 0, & \mathbf{x} \in \Omega_1(t) \\ 1, & \mathbf{x} \in \Omega_2(t) \end{cases} \tag{17.25}$$

(the characteristic function of the liquid phase). We therefore have the following corollary of 17C.

(17D) Thermodynamical laws for the quasi-static equations *Consider a solution of the quasi-static equations. Then for any two-phase control volume R,*

$$\frac{d}{dt}\int_R \varepsilon\,da = -\int_{\partial R} \mathbf{q}\cdot\mathbf{n}\,ds,$$

$$\frac{d}{dt}\int_R f(\theta)\,da \leq -\int_{\partial R} u\mathbf{q}\cdot\mathbf{n}\,ds + \int_{\partial \imath} \mathbf{C}\cdot\mathbf{v}_{\partial \imath}, \tag{17.26}$$

the difference between the right and left sides of $(17.26)_2$ being (17.19).

By (17.24), the rate of change of internal energy in the quasi-static theory has the simple form

$$\frac{d}{dt}\int_R \varepsilon\,da = -\text{area}(R_1)^{\cdot}. \tag{17.27}$$

17.3.4 Growth theorems

Problems arising from theories based on thermodynamics generally have associated global functionals (Lyapunov functions) that decrease on solution paths. As is clear from (15A), the same is true within the framework developed here, at least before the introduction of approximations. We now show that the (approximate) quasi-linear and quasi-static equations also have associated Lyapunov functions.

We restrict our attention to a *bounded* body and to the following types of boundary conditions.

1. Insulated boundary:

$$\mathbf{q}\cdot\mathbf{n} = 0 \quad \text{on } \partial\Omega \text{ for all time.} \tag{17.28}$$

2. Isothermal boundary:

$$u = U \quad \text{on } \partial\Omega \text{ for all time,} \tag{17.29}$$

with $U \equiv$ constant the prescribed **boundary temperature**.

For convenience, we write

$$\mathscr{F}(\mathfrak{s}) = \int_{\mathfrak{s}} f(\theta)\, ds \tag{17.30}$$

for the *free energy* of the interface at the transition temperature, and

$$\mathscr{D}(u,\mathfrak{s}) = \int_{\mathfrak{s}} b(\theta)V^2\, ds + \sum_{i=1,2}\int_{\Omega_i} \operatorname{grad} u \cdot \mathbf{K}_i \operatorname{grad} u\, da \geq 0 \tag{17.31}$$

for the total dissipation (17.19).

(17E) Growth theorem[91] *Let u be a solution of the quasi-linear equations with $c_1 = c_2 = c$.*

(a) *If the boundary is insulated,*

$$\left\{\operatorname{area}(\Omega_1) - c\int_\Omega u\, da\right\}^{\!\boldsymbol{\cdot}} = 0,$$

$$\left\{\mathscr{F}(\mathfrak{s}) + \tfrac{1}{2}c\int_\Omega u^2\, da\right\}^{\!\boldsymbol{\cdot}} = -\mathscr{D}(u,\mathfrak{s}) \leq 0. \tag{17.32}$$

(b) *If the boundary is isothermal,*

$$\left\{\mathscr{F}(\mathfrak{s}) + U\operatorname{area}(\Omega_1) + \tfrac{1}{2}c\int_\Omega (u-U)^2\, da\right\}^{\!\boldsymbol{\cdot}} = -\mathscr{D}(u,\mathfrak{s}) \leq 0. \tag{17.33}$$

Let u be a solution of the quasi-static equations.

(c) *If the boundary is insulated,*

$$\operatorname{area}(\Omega_1)^{\boldsymbol{\cdot}} = 0, \qquad \mathscr{F}(\mathfrak{s})^{\boldsymbol{\cdot}} = -\mathscr{D}(u,\mathfrak{s}) \leq 0. \tag{17.34}$$

---

[91] Cf. Gurtin (1986), Sections 10, 11; (1988), eqns (7.9), (7.10).

(d) *If the boundary is isothermal,*

$$\{\mathscr{F}(\mathfrak{s}) + U \text{ area}(\Omega_1)\}^{\cdot} = -\mathscr{D}(u, \mathfrak{s}) \leq 0. \tag{17.35}$$

PROOF. The relations (17.32) follow from (17.16), (17.18), (17.24), (17.28), and the fact that $\mathfrak{s}$ is a closed curve.

Next, for an isothermal boundary,

$$\int_{\partial\Omega} u\mathbf{q}\cdot\mathbf{n}\, ds = U \int_{\partial\Omega} \mathbf{q}\cdot\mathbf{n}\, ds; \tag{17.36}$$

(17.33) follows upon subtracting (17.18)$_1$ times U from (17.18)$_2$ and using (17.16), (17.24), (17.36), and the phrase following (17.18)$_2$.

Finally, (17.34) and (17.35) follow from (17.32) and (17.33) with $c = 0$. ∎

(17F) Remarks

1. In view of the agreement (17.11), for a solid–liquid system $\Omega_1(t)$ is the region occupied by the solid phase. If the boundary is supercooled, then $U < 0$ and U area$(\Omega_1) < 0$; in view of (17.33) and (17.35), *this at least indicates the tendency of the solid phase to grow.*

2. The result (17.34) provides a formal justification of the Wulff problem discussed in Section 7.6.

For an isotropic material the quasi-static equations reduce to the Mullins–Sekerka equations:[92]

$$\begin{aligned}\Delta u &= 0, & \mathbf{q} &= -k_i \text{ grad } u & &\text{in bulk,} \\ u &= -bV + fK, & V &= [\mathbf{q}]\cdot\mathbf{N} & &\text{on the interface,}\end{aligned} \tag{17.37}$$

with b, f, and k_i scalar constants and Δ the Laplacian. In this case (17.34) becomes

$$\text{area}(\Omega_1)^{\cdot} = 0, \qquad \text{length}(\mathfrak{s})^{\cdot} = -f^{-1}\mathscr{D}(u, \mathfrak{s}) \leq 0,$$

while (17.35) takes the form

$$\{f \text{ length }(\mathfrak{s}) + U \text{ area}(\Omega_1)\}^{\cdot} = -\mathscr{D}(u, \mathfrak{s}) \leq 0.$$

An analogous simplification holds for the quasi-linear problem.

---

[92] Mullins (1960), Mullins and Sekerka (1963, 1964), although these authors take $b = 0$. See also Yokoyama and Kuroda (1990), who use equations of this form with $f = 0$ and $b = b(\theta)$ to study pattern formation in the growth of snow crystals. The equations (17.37) also describe the motion of an interface separating immiscible viscous fluids, when the fluids lie in the narrow gap between parallel plates (Hele–Shaw cell) (cf. Saffman and Taylor (1958), McLean and Saffman (1981)).

17.3.5 Perfect conductors[93]

Consider the quasi-linear equations (17.13) for a bounded region with boundary held at the spatially constant temperature $U(t)$. We now discuss the asymptotic form these equations take *when the conductivity of each phase is large*. Precisely, we consider (17.13) with

$$\mathbf{K}_i \quad \text{replaced by} \quad \delta^{-1}\mathbf{K}_i \tag{17.38}$$

under the assumption that δ is small. Writing a formal perturbation for u in powers of δ, it is clear that the lowest-order term, also written u, should be consistent with

$$\begin{aligned} \text{div } \mathbf{q} = 0, \quad \mathbf{q} = -\mathbf{K}_i \text{ grad } u & \quad \text{in bulk,} \\ [\mathbf{q}] \cdot \mathbf{N} = 0 & \quad \text{on the interface,} \\ u = U & \quad \text{on } \partial\Omega, \end{aligned} \tag{17.39}$$

as well as the interface condition $u = g(\theta)K - b(\theta)V$. Under reasonable assumptions, the problem (17.39) has the unique solution $u(\mathbf{x}, t) \equiv U(t)$; the only equation then left to solve is the free-boundary condition

$$b(\theta)V = g(\theta)K - U(t) \quad \text{on the interface,} \tag{17.40}$$

which (for $U = $ constant) is the anisotropic evolution equation (8.6), (8.7) studied in Part I.[94]

(17G) Remarks

1. If, instead of a boundary at spatially uniform temperature, we consider an insulated boundary, then the condition $u = U$ in (17.39) is replaced by $[\mathbf{q}] \cdot \mathbf{n} = 0$ on $\partial\Omega$; (17.39) thus modified still has the solution $u(\mathbf{x}, t) \equiv U(t)$, but now $U(t)$ is *indeterminate*. On the other hand, for δ in (17.38) small but non-zero, $(17.32)_1$ is satisfied (granted $c_1 = c_2 = c$); we therefore expect that an *insulated perfect conductor* is described by the interface equations

$$b(\theta)V = g(\theta)K - U(t), \quad \text{area}(\Omega_1)^{\cdot} = c_0 U^{\cdot}, \tag{17.41}$$

   where $c_0 = c \text{ area}(\Omega)$.

2. For materials that have both *large conductivity* and *small specific heat*, one might consider the previous analysis with (17.38) supplemented by the replacement of c_i by δc_i. In this instance, the arguments leading to (17.40) remain unchanged, but (17.41) is replaced by

$$b(\theta)V = g(\theta)K - U(t), \quad \text{area}(\Omega_1)^{\cdot} = 0. \tag{17.42}$$

---

[93] Gurtin (1988).
[94] The scaling following (17.11) is essentially equivalent to setting $\ell = 1$; (5.10) and $(16.12)_1$ then demonstrate the consistency of (17.40) and (8.7). (Note the different definitions of $g(\theta)$ in (8.6) and (17.12).)

18
INSTABILITIES INDUCED BY SUPERCOOLING THE LIQUID PHASE[95]

Problems involving supercooled liquids and superheated solids are inherently unstable. In fact, for a crystal in a supercooled melt, it is precisely the unstable nature of the melt that induces growth of the crystal.[96] In this chapter we consider two simple examples which illustrate these instabilities; for simplicity, we limit our discussion to the isotropic quasi-static equations (17.37) with $b = 0$:

$$\Delta u = 0 \qquad \text{in bulk,}$$
$$u = fK, \quad V = [k_1(\text{grad } u)_1 - k_2(\text{grad } u)_2] \cdot \mathbf{N} \qquad \text{on the interface,}$$
(18.1)

where the subscripts 1 and 2 on grad u indicate the limit as the interface is approached from the corresponding phase. To ease the discussion, we will refer to phases 1 and 2 as the solid and the liquid, respectively, terminology that is justified by (17.11) (cf. Remark 16A).

18.1 The one-dimensional problem: growth of the solid phase from a seed of zero measure[97]

We first consider the one-dimensional problem corresponding to a flat interface with temperature variations only in the direction x normal to the interface. The body may then be identified with an interval, which, modulo a suitable scaling, we take to be $[0, 1]$. The temperature is then a function $u(x, t)$ and the interface may be identified with a point $\zeta(t)$ of $(0, 1)$. Assuming that the solid and liquid occupy the intervals $(0, \zeta(t))$ and $(\zeta(t), 1)$, respectively, the system (18.1) reduces to the differential equation

$$u_{xx}(x, t) = 0 \qquad x \in (0, 1), \quad x \neq \zeta(t), \quad t > 0, \qquad (18.2)$$

in conjunction with the free-boundary conditions

$$u(\xi(t), t) = 0 \qquad t > 0,$$
$$\zeta^{\cdot}(t) = k_1 u_x(\zeta(t) - 0, t) - k_2 u_x(\zeta(t) + 0, t) \qquad t > 0. \qquad (18.3)$$

---

[95] We confine our attention to situations in which the liquid is supercooled; a strictly analogous discussion applies when the solid is superheated.
[96] Cf., for example, Chalmers (1964); Delves (1974).
[97] Gurtin and Soner (1992). For the quasi-linear equations this problem is both more difficult and more interesting (cf. Sherman (1970); Friedman (1976); Fasano and Primicerio (1977); Fasano et al. (1981); Howison et al. (1985); Fasano et al. (1990)).

INSTABILITIES INDUCED BY SUPERCOOLING

The problem we will discuss concerns the *growth of a solid seed that initially occupies a single point*. Precisely, we consider the initial condition

$$\zeta(0) = 0, \tag{18.4}$$

so that initially the entire interval $(0, 1)$ is liquid, the solid being confined to the point $x = 0$. As boundary conditions, we assume that the end $x = 0$ is insulated, while the other end is held at the constant temperature U:

$$u_x(0, t) = 0, \quad u(1, t) = U \quad t > 0. \tag{18.5}$$

The underlying problem then consists of (18.2)–(18.5).

A solution of the differential equation (18.2) that satisfies the interface condition (18.3)$_1$ and the boundary condition (18.5) is

$$\begin{aligned} u(x, t) &= 0 & 0 \leq x \leq \zeta(t), \\ u(x, t) &= \frac{U[x - \zeta(t)]}{1 - \zeta(t)} & \zeta(t) < x \leq 1. \end{aligned} \tag{18.6}$$

Further, $u(x, t)$ satisfies the additional free-boundary condition (18.3)$_2$ if and only if $\zeta(t)$ satisfies the initial-value problem

$$\begin{aligned} \dot\zeta(t) &= -\frac{k_2 U}{1 - \zeta(t)}, \\ \zeta(0) &= 0. \end{aligned} \tag{18.7}$$

Thus for $U = 0$ the interface does not move, while $U > 0$ yields $\dot\zeta(0) < 0$, so that for $t > 0$ the body is entirely liquid at constant temperature T. Therefore *the solid seed cannot grow unless the liquid is supercooled*.

If

$$U < 0. \tag{18.8}$$

so that *the liquid is supercooled*, then (18.7) yields

$$\zeta(t) = 1 - \{1 - 2k_2|U|t\}^{\frac{1}{2}}. \tag{18.9}$$

For this solution the solid phase $(0, \zeta(t))$ forms spontaneously, since $\zeta(t) = 0$ at $t = 0$, and increases monotonically until $t = T = 1/2k_2|U|$, at which time the entire interval is solid. For $t \geq T$ this solution is not defined, but should be replaced by the solid equilibrium

$$u(x, t) \equiv U; \tag{18.10}$$

the lack of continuity of $u(x, t)$ in t at $t = T$ is of no concern, since the differential equation (18.2) does not involve derivatives with respect to time.

Exercise Obtain a solution for a body $\Omega = [-1, 1]$ consisting entirely of liquid except for a solid seed of zero measure placed at a point $\zeta_0 \in (-1, 1)$, assuming that the endpoints are equally supercooled: $u(-1, t) = u(1, t) = U < 0$.

18.2 Instability of a flat interface[98]

In this section we discuss infinitesimal perturbations of a flat interface moving at constant speed. This problem is important in two respects: it demonstrates the effects of capillarity; and it illustrates the unstable nature of the free-boundary problems under consideration.

Taking rectangular coordinates $\mathbf{x} = (x, y)$, we consider a flat interface moving with constant velocity V_0 in the y-direction. We consider a coordinate system that moves with the interface and is such that phase 1 occupies the region $\Omega_1 = \{y < 0\}$, and phase 2 the region $\Omega_2 = \{y > 0\}$. Then, letting u_1 and u_2, respectively, denote the restrictions of u to Ω_1 and Ω_2, the functions

$$u_1(x, y) = g_1 y, \qquad u_2(x, y) = g_2 y \qquad (18.11)$$

generate a solution of (18.1) with velocity V given by

$$V_0 = k_1 g_1 - k_2 g_2. \qquad (18.12)$$

The interfacial temperature corresponding to this solution is $u_1(x, 0) = u_2(x, 0) = 0$.

We now study the stability of this solution to small perturbations. We assume that the interface, relative to a coordinate frame moving with the underlying steady velocity V_0 of the unperturbed system, is represented as the graph of a function $y = h(x, t)$ with

$$h(x, t) = \delta(t) \sin \omega x. \qquad (18.13)$$

We assume that δ is small, so that the interface is flat to within a perturbation of order $O(\delta)$. Guided by (18.13), we consider temperature fields of the form

$$\begin{aligned} u_1(x, y, t) &= g_1 y + C_1(t)\, e^{\omega y} \sin \omega x, \\ u_2(x, y, t) &= g_2 y + C_2(t)\, e^{-\omega y} \sin \omega x; \end{aligned} \qquad (18.14)$$

these fields satisfy Laplace's equation and are asymptotic to the steady solution (18.11). Consider the free-boundary condition $u = fK$. At the interface $y = h(x, t)$ the fields (18.14) have the form

$$\begin{aligned} u_1(x, h(x, t), t) &= g_1 \delta(t) \sin \omega x + C_1(t)\, e^{\omega h(x, t)} \sin \omega x, \\ u_2(x, h(x, t), t) &= g_2 \delta(t) \sin \omega x + C_2(t)\, e^{-\omega h(x, t)} \sin \omega x, \end{aligned} \qquad (18.15)$$

while $(8.11)_3$ yields

$$K(x, t) = -\delta(t)\omega^2 \sin \omega x + O(\delta^2); \qquad (18.16)$$

the free-boundary condition $u = fK$ is therefore satisfied to within terms of

---

[98] Mullins and Sekerka (1964). Cf. Wagner (1956); Voronkov (1964). Cf. also the review articles of Sekerka (1968, 1973, 1984).

$O(\delta^2)$ by

$$C_1(t) = -[f\omega^2 + g_1]\delta(t), \qquad C_2(t) = -[f\omega^2 + g_2]\delta(t). \tag{18.17}$$

We have only to consider the remaining free-boundary condition. Since

$$\mathbf{N} = (1 + w^2)^{-\frac{1}{2}}(-w, 1), \qquad w = h_x, \tag{18.18}$$

grad $u_1 \cdot \mathbf{N}$ and grad $u_2 \cdot \mathbf{N}$, at the interface, are given by

$$\begin{aligned}g_1 - \delta(t)[f\omega^2 + g_1]\omega \sin \omega x + O(\delta^2), \\ g_2 + \delta(t)[f\omega^2 + g_2]\omega \sin \omega x + O(\delta^2),\end{aligned} \tag{18.19}$$

respectively. On the other hand, since we are using a coordinate frame moving with the steady velocity V_0, the perturbed velocity is, by virtue of $(8.11)_2$ and (18.13),

$$V(x, t) = V_0 + \delta^{\cdot}(t) \sin \omega x + O(\delta^2); \tag{18.20}$$

combining (18.19), (18.20), and the second interfacial condition in (18.1), and neglecting terms of order $O(\delta^2)$, we arrive at (18.12) and the differential equation

$$\begin{aligned}\delta^{\cdot} &= -\lambda(\omega)\delta, \\ \lambda(\omega) &= \omega\{f\omega^2(k_1 + k_2) + (k_1 g_1 + k_2 g_2)\}.\end{aligned} \tag{18.21}$$

Thus the interface is stable or unstable according as $\lambda(\omega) \geq 0$ or $\lambda(\omega) < 0$, with exponential growth of the perturbation for $\lambda(\omega) < 0$.

Assume that

$$g_1 \geq 0, \tag{18.22}$$

so that the temperature in the solid is at or below the transition temperature.

Case 1: $g_2 \geq 0$. Here *the liquid is not supercooled*, and, by (18.21), the interface is stable.

Case 2: $g_2 < 0$. Here *the liquid is supercooled*. If $|g_2|$ is small enough that $k_1 g_1 + k_2 g_2 > 0$, then the interface is stable; if $|g_2|$ is large enough that this sum is negative, then the interface is unstable for small values of ω, but stable for large values of ω. Perturbations with large ω and hence large curvature are stabilized by capillarity, as represented by f, but capillarity is not so important when ω is small, and here the interface is unstable.

Thus the interface is stable as long as the liquid is not supercooled, or even when the liquid is supercooled, provided the far-field temperature gradient is not too large. On the other hand, when the liquid is supercooled and the far-field temperature gradient is sufficiently large, the interface is unstable; in this case the unstable perturbations are those of large wavelength (small ω), as capillarity stabilizes perturbations of small wavelength.

REFERENCES

Abresch U. and Langer, J. (1986). The normalized curve shortening flow and homothetic solutions. *J. Diff. Geom.* **23**, 175–96.

Allen, S. M. and Cahn, J. W. (1979). A macroscopic theory for antiphase boundary motion and its application to antiphase domain coarsening. *Acta Metall.* **27**, 1085–95.

Almgren, F. J. (1966). *Plateau's Problem, an Invitation to Varifold Geometry.* Benjamin, New York.

Angenent, S. B. (1991). Parabolic equations for curves on surfaces – parts 1 and 2. *Ann. Math.* To appear.

Angenent, S. B. and Gurtin, M. E. (1989). Multiphase thermomechanics with interfacial structure, 2. Evolution of an isothermal interface. *Arch. Rational Mech. Anal.* **108**, 323–91.

Angenent, S. B. and Gurtin, M. E. (1992). Anisotropic motion of a phase interface: well posedness of the initial value problem and qualitative properties of the interface. To appear.

Barles, G. (1985). Remark on a flame propagation model. *Rapport INRIA* No. 464.

Brakke, K. A. (1978). *The Motion of a Surface by its Mean Curvature.* Princeton University Press.

Brush, L. N. and Sekerka, R. F. (1989). A numerical study of two-dimensional growth forms in the presence of anisotropic growth kinetics. *J. Crystal Growth* **96**, 419–41.

Cahn, J. W. and Hoffman, D. W. (1974). A vector thermodynamics for anisotropic surfaces – 2. Curved and faceted surfaces. *Acta Metall.* **22**, 1205–14.

Caroli, B., Caroli, C., and Roulet, B. (1984). Non-equilibrium thermodynamics of the solidification problem. *J. Crystal Growth* **66**, 575–85.

Chalmers, B. (1964). *Principles of Solidification.* Wiley, New York.

Chen, X.-F. and Reitich, F. (1992). Local existence and uniqueness of solutions of the Stefan problem with surface tension and kinetic undercooling. *J. Math. Anal. Appl.* **164**, 350–62.

Chen, Y.-G., Giga, Y. and Goto, S. (1991). Uniqueness and existence of viscosity solutions of generalized mean curvature flow equations. *J. Diff. Geom.* **33**, 749–86.

Chernov, A. A. (1963a). Crystal growth forms and their kinetic stability (in Russian). *Kristallografiya* **8**, 87–93. English trans. *Sov. Phys. Crystall.* **8**, 63–7.

Chernov, A. A. (1963b). Application of the method of characteristics to the theory of the growth forms of crystals (in Russian). *Kristallografiya* **8**, 499–505. English trans. *Sov. Phys. Crystall.* **8**, 401–5.

Chernov, A. A. (1971). Theory of the stability of face forms of crystals (in Russian). *Kristallografiya* **16**, 842–63. English trans. *Sov. Phys. Crystall.* **16**, 734–53.

Chernov, A.A. (1974). Stability of faceted shapes. *J. Crystal Growth* **24/25**, 11–31.

Coleman, B. D. and Mizel, V. J. (1963). Thermodynamics and departures from Fourier's law of heat conduction. *Arch. Rational Mech. Anal.* **13**, 245–61.

Coleman, B. D. and Noll, W. (1963). The thermodynamics of elastic materials with heat conduction and viscosity. *Arch. Rational Mech. Anal.* **13**, 167–78.

Crandall, M. G. and Lions, P.-L. (1983). Viscosity solutions of Hamilton–Jacobi equations. *Trans. Am. Math. Soc.* **277**, 1–42.

Dacorogna, B. and Pfister, C. E. (1991). Wulff theorem and best constant in Sobolev inequality. To appear.

Delves, R. T. (1974). Theory of interface instability. In *Crystal Growth* (ed. B. R. Pamplin). Pergamon, Oxford.

Di Carlo, A., Gurtin, M. E., and Podio-Guidugli, P. (1992). A regularized equation for anisotropic motion-by-curvature. *SIAM J. Math Anal.* **52**, 1111–19.

Dinghas, A. (1944). Uber einen geometrischen Satz von Wulff für die Gleichgewichtsform von Krystallen. *Zeit. Krystall.* **105**, 304–14.

Duchon, J. and Robert, R. (1984). Evolution d'une interface par capillarité et diffusion de volume, existence local en temps. *C. R. Acad. Sc. Paris*, **298**, Ser. 0, 473–6.

Evans, L. C. and Spruck, J. (1991). Motion of level sets by mean curvature 1. *J. Diff. Geom.* **33**, 635–81.

Evans, L. C., Soner, H. M., and Souganides, P. E. (1991). Phase transitions and generalized motion by mean curvature. *Comm. Pure Appl. Math.* To appear.

Fasano, A. and Primicerio, M. (1977). General free boundary problems for the heat equation. *J. Math. Anal. Appl.*; Part 1, **57**, 694–723; Part 2, **58**, 202–31; Part 3, **59**, 1–14.

Fasano, A., Primicerio, M., and Lacey, A. A. (1981). New results on some classical parabolic free boundary problems. *Quart. Appl. Math.* **38**, 439–60.

Fasano, A., Primicerio, M., Howison, S. D., and Ockendon, J. R. (1990). Some remarks on the regularization of supercooled one-phase Stefan problems in one dimension. *Quart. Appl. Math.* **48**, 153–68.

Fernandez-Diaz, J. and Williams, W. O. (1979). A generalized Stefan condition. *Zeit. Angew. Math. Phys.* **30**, 749–55.

Fonseca, I. (1989). Interfacial energy and the Maxwell rule. *Arch. Rational Mech. Anal.* **106**, 63–95.

Fonseca, I. (1991). The Wulff theorem revisited. *Proc. Roy. Soc. Lond.* A, **432**, 125–45.

Fonseca, I. and Müller, S. (1991). A uniqueness proof for the Wulff theorem. *Proc. Roy. Soc. Edin.* A, **119**, 125–36.

Frank, F. C. (1958). On the kinematic theory of crystal growth and dissolution processes. In *Growth and Perfection of Crystals* (ed. R. H. Doremus, B. W. Roberts, D. Turnbull), John Wiley, New York.

Frank, F. C. (1963). The geometrical thermodynamics of surfaces. In *Metal Surfaces: Structure, Energies, and Kinetics*, Am. Soc. Metals, Metals Park, Ohio.

Friedman, A. (1976). Analyticity of the free boundary for the Stefan problem. *Arch. Rational Mech. Anal.* **61**, 97–125.

Gage, M. (1984). Curve-shortening makes convex curves circular. *Inventiones Math.* **76**, 357–64.

Gage, M. (1986). On an area preserving evolution equation for plane curves. *Contemporary Math.* **51**, 51–62.

Gage, M. and Hamilton, R. S. (1986). The heat equation shrinking convex plane curves. *J. Diff. Geom.* **23**, 69–96.

Gjostein, N. A. (1963). Adsorption and surface energy (II): thermal faceting from minimization of surface energy. *Acta Metall.* **11**, 969–77.

REFERENCES

Grayson, M. A. (1987). The heat equation shrinks embedded plane curves to round points. *J. Diff. Geom.* **26**, 285–314.

Gurtin, M. E. (1986). On the two-phase Stefan problem with interfacial energy and entropy. *Arch. Rational Mech. Anal.* **96**, 199–241.

Gurtin, M. E. (1988). Multiphase thermomechanics with interfacial structure, 1. Heat conduction and the capillary balance law. *Arch. Rational Mech. Anal.* **103**, 195–221.

Gurtin, M. E. (1991). On thermomechanical laws for the motion of a phase interface. *Zeit. Angew. Math. Phys.* **42**, 370–88.

Gurtin, M. E. (1992). Thermodynamics and the supercritical Stefan equations with nucleations. *Quart. Appl. Math.* To appear.

Gurtin, M. E. and Soner, H. M. (1992). Some remarks on the Stefan problem with surface structure. *Quart. Appl. Math.* **52**, 291–303.

Gurtin, M. E. and Struthers, A. (1990). Multiphase thermomechanics with interfacial structure, 3. Evolving phase boundaries in the presence of bulk deformation. *Arch. Rational Mech. Anal.* **112**, 97–160.

Gurtin, M. E., Soner, H. M., and Souganides, P. E. (1992). Anisotropic motion of an interface relaxed by the formation of infinitesimal wrinkles. To appear.

Herring, C. (1951a). Surface tension as a motivation for sintering. In *The Physics of Powder Metallurgy* (ed. W. E. Kingston), McGraw-Hill, New York.

Herring, C. (1951b). Some theorems on the free energies of crystal surfaces. *Phys. Rev.* **82**, 87–93.

Hoffman, D. W. and Cahn, J. W. (1972). A vector thermodynamics for anisotropic surfaces – 1. Fundamentals and applications to plane surface junctions. *Surface Sci.* **31**, 368–88.

Howison, S. D., Ockendon, J. R., and Lacey, A. A. (1985). Singularity development in moving boundary problems. *Quart. Appl. Math.* **38**, 343–60.

Huisken, G. (1987). Deforming hypersurfaces of the sphere by their mean curvature. *Math. Z.* **198**, 134–46.

Ilmanen, T. (1991). Motion of level sets and of varifolds by mean curvature. Ph.D. Thesis, Dept. Math., University of California, Berkeley.

Langer, J. S. (1980). Instabilities and pattern formation in crystal growth. *Rev. Mod. Phys.* **52**, 1–27.

Luckhaus, S. (1990). Solutions for the two-phase Stefan problem with the Gibbs–Thomson law for the melting temperature. *Euro. J. Appl. Math.* **1**, 101–11.

McLean, J. W. and Saffman, P. G. (1981). The effect of surface tension on the shape of fingers in a Hele–Shaw cell. *J. Fluid Mech.* **102**, 455–69.

Moeckel, G. P. (1975). Thermodynamics of an interface. *Arch. Rational Mech. Anal.* **57**, 255–80.

Mullins, W. W. (1956). Two-dimensional motion of idealized grain boundaries. *J. Appl. Phys.* **27**, 900–4.

Mullins, W. W. (1960). Grain boundary grooving by volume diffusion. *Trans. Met. Soc. AIME* **218**, 354–61.

Mullins, W. W. and Sekerka, R. F. (1962). Application of linear programming theory to crystal faceting. *J. Phys. Chem. Solids* **23**, 801–3.

Mullins, W. W. and Sekerka, R. F. (1963). Morphological stability of a particle growing by diffusion or heat flow. *J. Appl. Phys.* **34**, 323–9.

Mullins, W. W. and Sekerka R. F. (1964). Stability of a planar interface during solidification of a dilute binary alloy. *J. Appl. Phys.* **35**, 444–51.

Osher, S. and Sethian, J. A. (1988). Front propagation with curvature dependent speed: Algorithms based on Hamilton–Jacobi formulations. *J. Comput. Phys.* **79**, 12–49.

Protter, M. and Weinberger, H. (1967). *Maximum Principles in Differential Equations.* Prentice-Hall, New York.

Rogers, J. C. W. (1983). The Stefan problem with surface tension. In *Free Boundary Problems: Theory and Applications* (ed. A. Fasano and M. Primicerio). Pitman, Boston.

Rubinstein, J., Sternberg, P., and Keller, J. B. (1989). Fast reaction, slow diffusion, and curve shortening. *SIAM J. Appl. Math.* **49**, 116–33.

Saffman, P. G. and Taylor, G. I. (1958). The penetration of a fluid into porous medium or Hele–Shaw cell containing a more viscous liquid. *Proc. Roy. Soc. Lond.* **A245**, 312–29.

Seidensticker, R. G. (1966). Stability considerations in temperature gradient zone melting. In *Crystal Growth* (ed. H. S. Peiser). Pergamon, Oxford.

Sekerka, R. F. (1968). Morphological stability. *J. Crystal Growth* 3, 4, 71–81.

Sekerka, R. F. (1973). Morphological stability. In *Crystal Growth: an Introduction.* North-Holland, Amsterdam.

Sekerka, R. F. (1984). Morphological instabilities during phase transformations. In *Phase Transformations and Material Instabilities in Solids* (ed. M. E. Gurtin). Academic Press, New York.

Sethian, J. A. (1985). Curvature and the evolution of fronts. *Comm. Math. Phys.* **101**, 487–99.

Sherman, B. (1970). A general one-phase Stefan problem. *Quart. Appl. Math.* **28**, 377–82.

Soner, H. M. (1990). Motion of a set by the curvature of its boundary. *J. Diff. Eqns.* To appear.

Tarshis, L. A. and Tiller, W. A. (1967). The effect of interface-attachment kinetics on the morphological stability of a planar interface during solidification. In *Crystal Growth* (ed. H. S. Peiser). Pergamon, Oxford.

Taylor, J. E. (1978). Crystalline variational problems. *Bull. Am. Math. Soc.* **84**, 568–88.

Taylor, J. E. (1988). Constructions and conjectures in crystalline nondifferential geometry. *Proceedings of the Conference on Differential Geometry*, Rio de Janeiro, Pitman, London.

Taylor, J. E. (1991). Motion of curves by crystalline curvature, including triple junctions and boundary points. To appear.

Truesdell, C. and Noll, W. (1965). The non-linear field theories of mechanics. In *Handbuch der Physik*, Vol. III (ed. S. Flugge). Springer-Verlag, Berlin.

Visintin, A. (1988). Surface tension effects in phase transition. In *Material Instabilities in Continuum Mechanics* (ed. J. M. Ball). Clarendon Press, Oxford.

Visintin, A. (1989). Stefan problem with surface tension. In *Mathematical Models for Phase Change Problems* (ed. J. F. Rodrigues). Birkhäuser Verlag, Basel.

Voronkov, V. V. (1964). Conditions for formation of mosaic structure on a crystallization front (in Russian). *Fizika Tverdogo Tela* **6**, 2984–8. English trans. *Sov. Phys. Solid State* **6**, 2378–81.

Wollkind, D. J. (1979). A deterministic continuum mechanical approach to morphological stability of the solid-liquid interface. In *Preparation of Properties of Solid State Materials* (ed. C. R. Wilcox). Dekker, New York.

Wulff, G. (1901). Zur Frage der Geschwindigkeit des Wachsthums und der Auflösung der Krystallflachen. *Zeit. Krystall. Min.* **34**, 449–530.

Yokoyama, E. and Kuroda, T. (1990). Pattern formation in growth of snow crystals occurring in the surface kinetic process and the diffusion process. *Physical Review A*, **41**, 2038–49.

Young, L. C. (1969). *Lectures on the Calculus of Variations and Optimal Control Theory*. W. B. Saunders, Philadelphia.

INDEX

angle-set 7
arc-length
  description 14
  map 6, 14, 21
  trajectory 15
area identity 10

balance of energy 112, 119
balance of forces 32, 33, 105
boundary conditions, thermodynamical theory 135
boundary curve 7
bounded curve 6

capillary force 32, 40, 93, 132
closed curve 6
compatibility theorem
  mechanical theory 38
  regularized theory 108
  thermodynamical theory 115, 126
constitutive equations
  mechanical theory 38
  regularized theory 107
  thermodynamical theory 114, 124, 126
containment principle 75, 99
control volume 26–30
convex curve 7–10
convexification 42
convexified energy 46
convexity 42
corners 40, 94
corner force 94
crystalline motions 96
curvature 7
curve
  class R 8
  definition 6
  normal, tangent for 6, 7
  parametrization of 6
curve-shortening equation 1

dissipation inequality
  mechanical theory 35
  regularized theory 106

energy, bulk 35
energy, interfacial
  crystalline 91
  definition 35
  nonsmooth 91
  regular 51
  stability of 50
  total 70
  unstable 78
entropy
  bulk 112
  interfacial 119
entropy production
  bulk 113
  interfacial 123
equilibrium equations 53
evolving curve 14
extended energy 42, 45

faceted curve 11, 22, 81
first law 112
Frank
  diagram 42, 44
  potential 44
free energy
  bulk 114
  interfacial 120

Gibbs relation 127
globally convex section 42
globally stable angles and sections 50
growth of entropy 112, 119
growth theorem, thermodynamical theory 135

INDEX

heat flux, bulk 112
heat flux, apparent 119
heat supply, bulk 112
heat equation 117

infinitesimally wrinkled curve 12–13, 85, 88, 100
interface
  conditions for thermodynamical theory 130, 131
  evolving 14
  a level set 62
  stationary 64
  steadily evolving 19, 65–68, 90
  the graph of a function 62, 84
internal energy
  bulk 112
  interfacial 119

jump in a function 27
juncture
  angle 21
  curvature 21
  definition of 11
  tangential endpoint velocities of 21

kinetic coefficient 39, 86

latent heat 125
localization lemma 30

Mullins–Sekerka equations 136, 140

normal angle 7
normal interaction 36
normal time derivative 15
normal velocity 15

perfect conductors 137
phase region 26, 118
polar diagram 41
power 33–34
PS curve 11
PS evolving curve
  admissibility for 78
  definition 20

quasi-linear problem 131
quasi-static problem 132

reference region 7
regular interfacial energy 51
regularized theory 105

second law
  general discussion 1
  mechanical theory 35
  thermodynamical theory 112
simple curve 6
Soner's solution 75
spinodal decomposition 110
stability
  local 49
  for a PS curve 47
steadily evolving
  bump 20
  facet 20
supercooling, superheating 125, 139, 141
support function 8
surface shear 32
surface tension 32

Taylor's solution 99
temperature 112
thermodynamical laws, approximate 132, 134
transition temperature 125
transport theorem
  area and perimeter 18
  integrals 18, 23

unbounded curve 6

variational lemmas 24–5

wrinkled curve 11, 22, 81
Wulff
  problem 58
  ratio 58
  region 55, 92
  Theorem 58